Challenging Obesity
The science behind the issues

Challenging Obesity

The science behind the issues

Edited by Heather McLannahan and Pete Clifton

The Open University

OXFORD
UNIVERSITY PRESS

Published by Oxford University Press, Great Clarendon Street, Oxford OX2 6DP
in association with The Open University, Walton Hall, Milton Keynes MK7 6AA.

Oxford University Press is a department of the University of Oxford. It furthers the University's
objective of excellence in research, scholarship, and education by publishing worldwide in

Oxford New York

Auckland Cape Town Dar es Salaam Hong Kong Karachi Kuala Lumpur Madrid Melbourne
Mexico City Nairobi New Delhi Shanghai Taipei Toronto

with offices in
Argentina Austria Brazil Chile Czech Republic France Greece Guatemala Hungary
Italy Japan Poland Portugal Singapore South Korea Switzerland
Thailand Turkey Ukraine Vietnam

Oxford is a registered trade mark of Oxford University Press in the UK and in certain
other countries.

Published in the United States by Oxford University Press Inc., New York

Edited and designed by The Open University.

Typeset by SR Nova Pvt. Ltd, Bangalore, India.

Printed and bound in the United Kingdom by the University Press, Cambridge.

This book forms part of the Open University course SDK122 *Challenging obesity*. Details of this
and other Open University courses can be obtained from the Student Registration and Enquiry
Service, The Open University, PO Box 197, Milton Keynes MK7 6BJ,
United Kingdom:
tel. +44 (0)870 333 4340, email general-enquiries@open.ac.uk.

http://www.open.ac.uk

British Library Cataloguing in Publication Data available on request

Library of Congress Cataloging in Publication Data available on request

ISBN 978 0 19 956337 1

10 9 8 7 6 5 4 3 2 1

ABOUT THIS BOOK

This book forms part of an Open University (OU) course for students studying an undergraduate programme in Health Sciences. The book has also been designed to 'stand alone' for readers studying it in isolation from the rest of the course, either as part of an educational programme at another institution, or for general interest and self-directed study.

Challenging Obesity is a multidisciplinary introduction to a topic of global importance for public health. This book is for anyone seeking a scientific understanding of obesity. In such a wide-ranging subject area, we have had to be selective, but we have integrated aspects of the biology and psychology with health statistics and social studies to illuminate the causes for people becoming overweight and obese and developing associated diseases, the impacts of these conditions on societies and individuals and the science underlying common treatments. No previous experience of studying science has been assumed and new concepts and specialist terminology are explained with examples and illustrations.

Our intention is to bring you into the subject, develop your confidence through activities and guidance, and provide a stepping stone into further study. The most important terms appear in **bold** in the text at the point where they are first defined, and these terms are also in bold in the index at the end of the book. Understanding of the meaning and uses of these terms is essential (i.e. assessable) if you are an OU student. At various points in the book, you will find 'boxed' material. Boxes contain two types of information: basic concepts explained in the kind of detail that someone who is completely new to the health sciences is likely to want; and also some advanced concepts that will enrich your understanding further, should you wish to delve deeper into the subject matter.

Active engagement with the material throughout this book is encouraged by numerous 'in text' questions, indicated by a diamond symbol (◆) followed immediately by our suggested answers. It is good practice always to cover the answer and attempt your own response to the question before reading ours. Some chapters contain activities to help you develop and practice particular skills. At the end of each chapter, there is a summary of the key points and a list of the main learning outcomes, followed by self-assessment questions to enable you to test your own learning. The answers to these questions and activities are at the end of the book – where you will also find the acknowledgements, which give the sources of data for many of the figures and tables.

Internet database (ROUTES)

A large amount of valuable information is available via the internet. To help OU students and other readers of this book to access good quality sites without having to search for hours, the OU has developed a collection of internet resources on a searchable database called ROUTES. All websites included in the database are selected by academic staff or subject-specialist librarians. The content of each website is evaluated to ensure that it is accurate, well presented and regularly updated. A description is included for each of the resources.

The website address for ROUTES is: http://routes.open.ac.uk/

Entering the Open University course code 'SDK122' in the search box will retrieve all the resources that have been recommended for this book. Alternatively, if you want to search for any resources on a particular subject, type in the words which best describe the subject you are interested in (for example, 'bariatric surgery'), or browse the alphabetical list of subjects.

Authors' acknowledgements

As ever in The Open University, this book combines the efforts of many people with specialist skills and knowledge in different disciplines. At the outset, discussions with Dr Bryony Butland and colleagues from the UK Government's 'Tackling Obesities' project were helpful. The OU authors, Heather McLannahan, Vicky Taylor and Terry Whatson, who were joined by Pete Clifton (Professor of Psychology, University of Sussex) and Ali Prust (General Practitioner, Isle of Jura), would like to acknowledge the critical comments of their colleagues Joe Buchanan (Editor) and Claire Rothwell (Science) and of Ann-Louise Dyer, all of whom made numerous comments and suggestions for improvements. We are very grateful to our External Assessor, Professor Peter Kopelman, Principal, St George's, University of London, whose detailed comments have contributed to the structure and content of the book and kept the needs of our intended readership to the fore.

Special thanks are due to all those involved in the OU production process, chief among them Viki Burnage, our wonderful Course Manager, whose commitment, efficiency and unflagging good humour were at the heart of the endeavour, and who was ably assisted by Course Team Assistants Becky Efthimiou, Pat McVity, Dawn Partner and Yvonne Royals. We also warmly acknowledge the contributions of Steve Best, our graphic artist, who developed and drew all the diagrams; Sarah Hofton and Chris Hough, our graphic designers, who devised the page designs, layouts and cover; and Martin Keeling, who carried out picture research and rights clearance. The media project manager was Judith Pickering.

For the copublication process, we would especially like to thank Jonathan Crowe of Oxford University Press and, from within The Open University, Christianne Bailey (Media Developer, Copublishing). As is the custom, any small errors or shortcomings that have slipped in (despite our collective best efforts) remain the responsibility of the authors. We would be pleased to receive feedback on the book (favourable or otherwise). Please write to the address below.

Dr Heather McLannahan, SDK122 Course Team Chair

Department of Life Sciences
The Open University
Walton Hall
Milton Keynes
MK7 6AA
United Kingdom

Environmental statement

Paper and board used in this publication is FSC certified.

Forestry Stewardship Council (FSC) is an independent certification, which certifies that the virgin pulp used to make the paper/board comes from traceable and sustainable sources from well-managed forests.

CONTENTS

CHAPTER 1

A GLOBAL CHALLENGE

We're just too darned fat, ladies and gentlemen … Our poor eating habits and lack of activity are literally killing us and they're killing us at record levels.

Tommy Thompson, US politician, March 2004

1.1   Introduction

In words that everyone could understand, Tommy Thompson, Secretary of the Health and Human Services Department, reported the fact that poor diet and physical inactivity were poised to become the major cause of preventable deaths in the USA. The number of deaths (**mortality**) caused by these two factors had risen by a third (33%) over the previous 10 years. These 365 000 deaths represented 15% of deaths from all causes and an estimated US$17 billion cost to the US economy. As the new century dawned, 64% of the US adult population were 'too darned fat'.

A billion is a thousand million.

The scale of the problem, which has become known as the 'obesity epidemic', was worrying other countries too and in October 2007 the UK Government launched its 'Tackling Obesities' project and reported that most adults in the UK were already overweight, and predicted that by 2050 around 60% of men and 50% of women would be clinically obese. The criteria for deciding whether an adult is overweight and the definition of terms such as 'clinically obese' will be explained in Section 1.3, but a common-sense approach might suggest that obesity could be considered an appropriate description of when being 'too darned fat' starts to interfere with everyday living. Further thought might indicate that obesity must have an association with long-term health prospects. Indeed, this is the case and, as with the report from the USA introduced by Tommy Thompson, the cost of the associated health care was a concern raised by the UK 'Tackling Obesities' project. It was suggested that, unless action was taken, obesity-related diseases would cost an extra £45.5 billion per year by 2050.

This book examines a wide range of aspects of obesity, including social and cultural factors, with a particular emphasis on biological and psychological factors in order to understand how to prevent and treat obesity in the future.

As statistics from the World Health Organization (WHO) show in Figures 1.1 and 1.2, the USA and the UK are not alone in facing this problem.

Figures 1.1 and 1.2 give the **prevalence** of overweight and obesity in several countries in 2005. Prevalence is the total number of people who have a condition at a particular point in time, regardless of how long they have been affected. It is often expressed as a *rate*, i.e. the number of cases per 1000 or per 100 people in the population in question. When expressed as the number of cases per 100, it is known as the percentage (%).

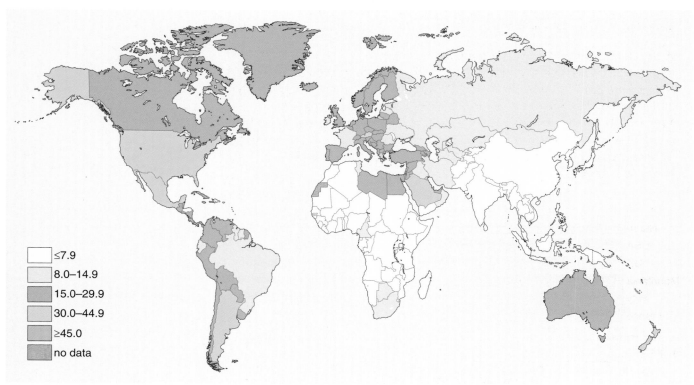

(a) Estimated obesity prevalence in men aged 30 or over in 2005/%

Legend:
- ≤7.9
- 8.0–14.9
- 15.0–29.9
- 30.0–44.9
- ≥45.0
- no data

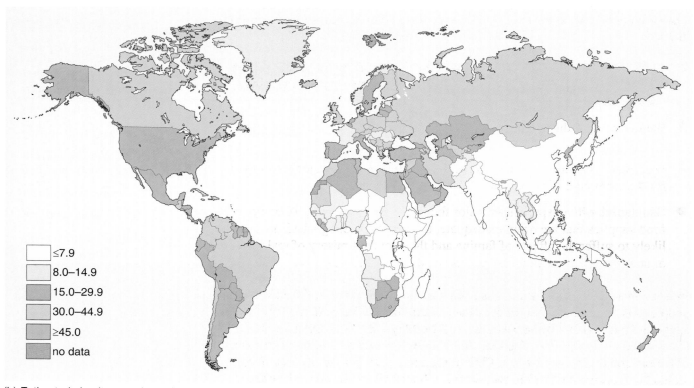

(b) Estimated obesity prevalence in women aged 30 or over in 2005/%

Legend:
- ≤7.9
- 8.0–14.9
- 15.0–29.9
- 30.0–44.9
- ≥45.0
- no data

Figure 1.1   Estimated prevalence of obesity across the world, expressed as a percentage, in (a) men and (b) women, aged 30 or over in 2005. ≤, less than or equal to; ≥, greater than or equal to.

Box 1.1 Reading a bar chart

A bar chart is a simple way of presenting numerical data visually, so as to emphasise the relative size of different numbers. In the example below, the bars are drawn *horizontally*, with the longest at the top and the shortest at the bottom, but bar charts can also have the bars drawn *vertically* (as in Figures 1.16 and 1.18 later in this chapter). In Figure 1.2, almost 75% of the adult female population of the USA is overweight or obese, which is nearly 3 times the comparative figure for China, with 25%, so the bar for the USA is nearly 3 times the length of the bar for China.

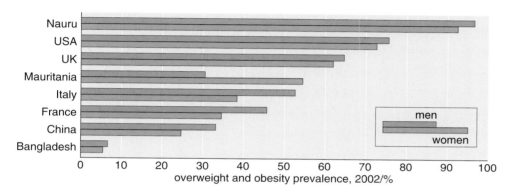

Figure 1.2 Estimates of the prevalence of overweight and obesity, expressed as a percentage of that population, in people aged 15 or over from several countries in 2005.

◆ Look at Figure 1.2 and decide which country has the lowest prevalence of overweight and obese people. What is the prevalence of overweight and obesity in that country?

◆ Bangladesh has the lowest prevalence, at around 6%.

◆ From your general knowledge, suggest why this figure is much lower than for the other countries in Figure 1.2.

◆ Bangladesh often experiences major food shortages following flooding. With food supplies that are often inadequate, the people of Bangladesh are more likely to suffer the effects of famine and the attendant misery of malnutrition or starvation than to face the opposite challenge of obesity.

Indeed, obtaining enough to eat has been a major preoccupation for most people most of the time that humans have inhabited the Earth. It has even been suggested that whole civilisations, such as the Maya of South America, may have been wiped out by famine. The idea that a time might come when vast numbers of people could have access to sufficient food to allow them to become obese would have been consigned to the realms of fantasy by most of our forebears. Most would have been in sympathy with the views of the English mathematician and philosopher Thomas Malthus (1766–1834), who predicted that unchecked increase in populations would lead to widespread famine and death.

Population increase has happened, but there are more overfed than malnourished people in today's overcrowded world. Current predictions are that obesity prevalence will rise and it is obesity that will become a leading cause of death, rather than famine. These predictions are based on trends from those countries that have reliable records from local and national surveys. Within Europe, surveys suggest that over 20% of adults are obese and another 30% are overweight, so over 50% of the population are affected overall. Figure 1.3 shows the estimated prevalence of adult obesity in Europe from 1985–2005 and the predicted levels of obesity to 2050.

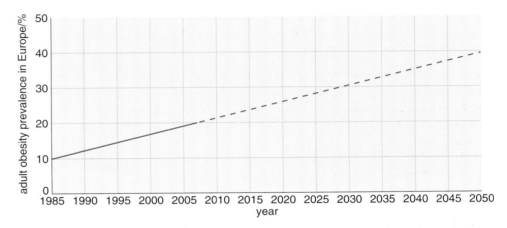

Figure 1.3   Estimated prevalence of adult obesity in Europe from 1985–2005 and projected levels of obesity to 2050.

◆ What, according to Figure 1.3, was the prevalence of adult obesity in Europe in 1985? By which year had it doubled?

◆ In 1985, the prevalence was 10%. It had doubled to 20% by 2007.

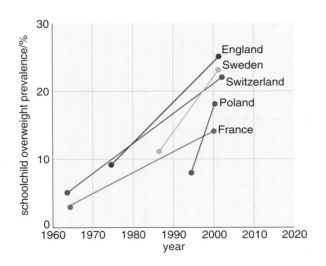

Figure 1.4   Prevalence of overweight in schoolchildren in selected European countries from 1960–2000.

The prediction that the prevalence of obesity will continue to increase is in part based on the statistics for childhood obesity (see Section 1.3.2). The number of children who are overweight has increased in all of the European countries surveyed. Figure 1.4 shows results from a few of these countries. Obese children are likely to become obese adults, as you will read in Chapter 6. There is evidence that the longer a person has been obese, the greater their risk of premature death.

The connection between obesity and the risk of ill health has been appreciated in Western cultures for at least two thousand years, and was neatly expressed by the Greek historian and writer Plutarch (AD 46–122) when he observed that 'thin people are generally the most healthy; we should not therefore indulge our appetites with delicacies or high living, for fear of growing corpulent' (Mackenzie, 1758, cited by Haslam, 2007). Obesity first became a noticeable health issue in the USA and was first recognised as a **chronic disease** (a long-lasting disease, usually with a gradual onset) in 1985 by the USA's National Institutes of Health, but there are many people who would disagree with this classification. While it is an acknowledged public health threat, some people prefer to consider obesity as a **risk factor** for other diseases rather than as a disease in its own right. A risk factor is any factor which is *associated* with an increased chance of developing a particular disease, for an individual or for a population. It is the frequent associations found between obesity and serious diseases such as diabetes mellitus – type 2 – and cardiovascular diseases that has led to obesity being cited as a major killer. Similarly, the smoking of tobacco is a risk factor for many diseases and is cited as the number one cause of preventable deaths; however, smoking is not classified as a disease. A number of obese people assert that, just as not every smoker dies of lung cancer, not all people who are obese are ill and at risk of dying from a related illness.

It can also be argued that referring to obesity as a disease and talking of an obesity epidemic places the issue in the biomedical arena, allowing society to await a 'cure' rather than take responsibility for seeking wide-ranging, non-medical, preventative solutions to the problem. We can draw a further parallel with smoking. Although smoking-related deaths are still, at the time of writing (2007), the leading cause of avoidable death in countries such as the UK and the USA, their **incidence** (the number of new cases in a particular period, usually per year and usually expressed as a rate) has been declining. Factors that have contributed to the decline in smoking include the provision of information through government-backed campaigns and education. Taxes and legislation have also played an important part. Whether such a strategy can work for obesity will be discussed later in this book.

This chapter has used – and will continue to use – information drawn from a branch of the health sciences called **epidemiology**. Epidemiology is the statistical study of the occurrence, distribution, potential causes and control of diseases and disabilities in populations. In the rest of this chapter, you will find out how obesity is defined and measured, and how the health of different groups of people is affected by obesity – as well as considering the question of how and why obesity has come to be so prevalent. In later chapters, you will explore these themes in greater detail and consider what kinds of actions might reverse the trend.

1.2 What is obesity?

Being overweight and being obese are not one and the same thing. However, both overweight and obesity classifications are associated with increased incidence of serious conditions such as hypertension (high blood pressure), cardiovascular disease and diabetes, as well as conditions that are not life-threatening, but which do affect wellbeing, such as varicose veins and arthritis. Everyone needs some body fat, for reasons that you will read about later, but obesity is different. **Obesity** has been defined as having an excess accumulation of body fat (Prentice and Jebb, 2003). It is the excess body fat that creates the most serious health problems – although carrying around extra weight creates some practical problems too (see Vignette 1.1). There will be more about the effects of additional body weight on body functions in Chapter 7.

Type 2 diabetes (also known as mature-onset diabetes) will be explained in Chapter 7. It is distinct from type 1 diabetes. However, in the rest of this book, we will use the term diabetes to refer to type 2 diabetes, and diabetes mellitus when we are referring to both types.

Cardiovascular diseases are diseases of the heart and blood vessels.

It should be noted that death itself is not avoidable! However, if longevity (length of life) is being curtailed then death is described as avoidable or preventable.

Vignette 1.1   Introducing Charlie

Charlie is 12 years old and, although he would never admit it, he enjoys school. The school is in a small town where Charlie lives and it takes him about 10 minutes to walk there. He has many friends, most of whom live locally and all meet up as they walk to school. Others are bussed in from surrounding villages or dropped off by their parents as they go to work. Charlie has always been a little on the tubby side, but he's not alone: there are lots of kids fatter than him.

Football is a great love of his. He is an enthusiastic Arsenal supporter and for his last birthday his parents took him and three of his special friends to an Arsenal match. The boys play football at school and two of his friends are in the school under-13s team. This means Charlie no longer plays as frequently as they do and he finds that he gets rather out of breath when they all kick a ball around together. He also notices that he doesn't like having to go upstairs to collect stuff from his room, so he tries to keep organised so he doesn't have to rush back upstairs to collect things like his homework just before he sets off for school.

◈   This vignette suggests that Charlie is carrying a bit too much weight. What practical problems has he experienced and what measures has he taken to overcome them?

◆   Charlie gets out of breath when he runs around and finds going up stairs a bit of an effort. So he has adapted his lifestyle to avoid getting out of breath, which leads to a more inactive life.

You'll read more about Charlie later in this chapter.

Fat is stored in **adipocytes** (fat cells; Figure 1.5b). Our bodies are composed of millions of cells – too small to be seen with the unaided eye. Cells have one basic design (Figure 1.5a), but each type of cell is specialised according to its function. For example, **neurons** (nerve cells; Figure 1.5c) are the communication channels within the brain and between the brain and all other parts of the body, and gut cells (Figure 1.5d) absorb nutrients from food matter moving through the gut (about which you will learn more in Chapter 3). Adipocytes are relatively large, clear cells that can alter in volume, perhaps as much as tenfold, as adult humans become more obese or less obese. The number of adipocytes in an adult is influenced by nutrition in childhood and even before that, i.e. while in the womb (see Chapter 6).

Cells with a common function are grouped together to form tissues. **Adipose tissue** (fat tissue) comprises about 80% adipocytes with other cells, and material secreted by them, binding the adipocytes together. It has a blood supply and a nerve supply, and both these are also considered to be part of the adipose tissue. Adipose tissue is concentrated into characteristic areas as *fat depots*. These are found beneath the skin (subcutaneous; hence the advice to avoid chicken skin if on a low-fat diet), within the muscles (intramuscular; this can be seen as the

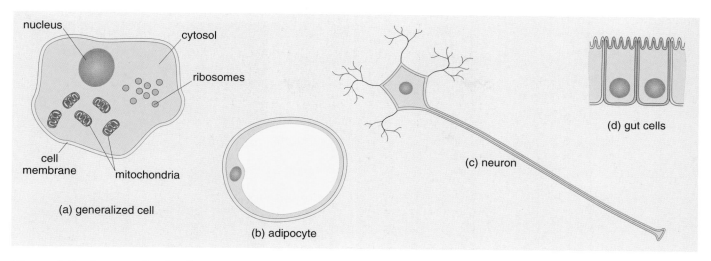

Figure 1.5 A generalised cell and cells with a specific function, reflected in their *morphology* (shape). (a) Generalised structure of a cell, showing the cell membrane and the watery cytosol in which can be seen some mitochondria and ribosomes. The other cell structure shown is the nucleus; this is where chromosomes containing the genetic material are housed. Other cell structures exist but are not shown in this diagram. (b) Adipocytes are essentially spherical, allowing them to expand as more fat (shown in yellow) is stored and contract as circumstances dictate much in the manner of a balloon. (c) Some neurons resemble branching trees. This shape increases their surface area and allows them to receive information from many neighbouring cells. (d) The surface of the cells lining the gut have their surface area increased to maximise their absorptive surface.

white 'marbling' in a piece of steak) and **visceral fat** (intra-abdominal fat). When the visceral fat depot becomes excessively large, it is known scientifically as *central obesity* but in everyday speech is often referred to as a 'beer belly'. It is the increase in size of this depot in particular that gives rise to concerns for ill health. The principal fat depots are also characteristic of gender, as shown in Figure 1.6a and b (overleaf). Women's breasts are composed almost entirely of adipose tissue; together with the fat around thighs and buttocks, these are responsible for the desirable, curvaceous figure beloved of artists such as Rubens (Figure 1.6d). The paunch, characteristic of many men, is formed from a thickening of the outer wall of the abdomen. These different fat depots show differential tendencies to increase in size as a person becomes more obese. For example, in men the paunch and subcutaneous fat in the thigh may be greatly increased, while the intramuscular fat in the calf hardly changes at all (Figure 1.6c).

Adipose tissue has a number of functions. You might (correctly) think of adipose tissue as a useful store that can be raided during times of food shortage. It is also known that the release of fats from adipose tissue is an important source of energy during intense or protracted exercise. Interestingly, it turns out that there are substantial differences in the ability of different fat depots to release fats at such times (see Chapter 8). Much of the release comes from upper-body rather than lower-body subcutaneous depots. This can come as a disappointment to those who join a gym to get 'into shape'!

It is also increasingly clear that adipose tissue plays a critical role in regulating some of the processes involved in eating and using food. Many of these processes are complex and some are, as yet, not fully understood. They will be the subject of later chapters.

Figure 1.6 Fat depots in (a) men and (b) women. These depots become excessively large in (c) obese men, as seen in Rubens's *Bacchus* of 1638 and in (d) obese women, as seen in Rubens's *Venus at a Mirror* of 1615.

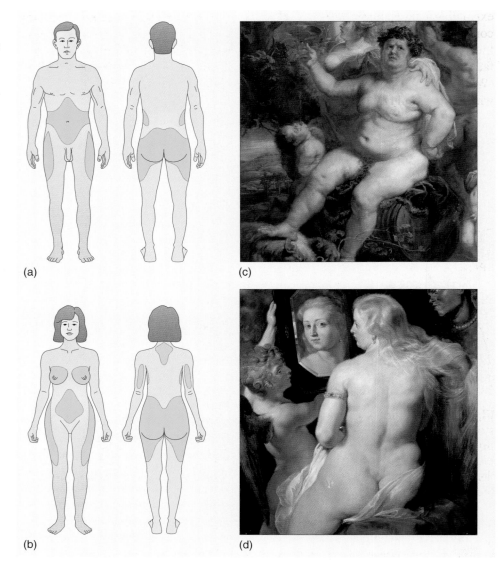

(a)

(b)

(c)

(d)

One of the means by which adipose tissue influences other areas of bodily activity is through the release of **hormones**. Hormones are chemicals that have a signalling function; they are released by the cells of one tissue, called an **endocrine gland**, and are then carried around the body in the bloodstream. When they pass by cells that have *receptors* which can interact with and bind to the hormone, the hormone will be taken out of the bloodstream and will bind to cells in this *target tissue*. In summary: hormones are released by one tissue (such tissues are called endocrine glands), are carried by the blood and have their effect on a distant target tissue, or tissues. Adipose tissue is an endocrine gland. It is also a target tissue interacting with hormones released from other endocrine glands (see Figure 1.7), as you will read in Chapter 4. Hormones *released* by adipocytes affect aspects of food-related behaviour such as appetite, as well as storage of fats and sugars. If the amount of fat in a depot increases, it will alter the amount and sometimes the type of hormones that are released. The effect of one depot expanding – the fat around the heart, say – can have a different outcome from the effect of a different depot expanding – for example, the hips. In this way, obesity can influence how the body consumes and uses food in varied, far-reaching and, unfortunately, negative ways. Indeed, an alternative definition of obesity is 'a condition of abnormal or excessive fat accumulation in adipose tissue, to the extent that health may be impaired' (WHO, 2002). Thus obesity can be defined by the quantity of adipose tissue and where it is found but not by body weight.

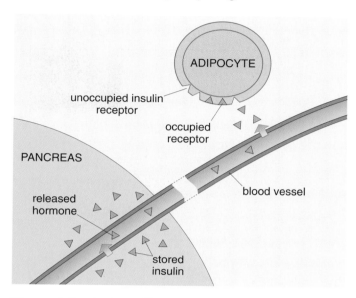

Figure 1.7   Action of hormones on receptor cells. Here, the hormone insulin is stored in the pancreas (the endocrine gland in this example; see also Chapter 3). Some molecules of insulin are released into the blood vessel and travel in the blood. When they pass the adipose tissue (the target tissue), the hormone leaves the blood vessel and binds to available (unoccupied) receptors on adipocytes (cells in the target tissue).

1.2.1   Summary

People who are obese have an increased risk of developing other serious illnesses – especially diabetes and cardiovascular diseases. Obesity can therefore best be defined as an excess accumulation of body fat that predisposes to ill health.

Fat is stored in the adipocytes of adipose tissue. Adipose tissue is found in gender-specific sites. Fat depots are also endocrine organs, releasing different hormones according to circumstances, as will be discussed in Chapter 4.

1.3   Measuring obesity

1.3.1   BMI in adults

Although obesity is not defined by weight, it is the case that most people's weight reflects the amount of fat they are carrying. In general conversation, you are more likely to hear someone bemoaning the fact that they have 'put on weight' rather than that their clothes seem to be shrinking! However, while a weight of 80 kilograms (12 stone 7 lb) would describe an obese person of 1.6 metres (5 ft 2 in) tall, the same weight would be a healthy weight for someone

Approximate conversion factors:
1 stone = 6.35 kilograms (kg)
1 pound (lb) = 0.45 kg
1 foot (ft) = 0.305 metres (m)
　　　　　 = 30.5 centimetres (cm)
1 inch (in) = 2.5 cm

1.8 metres (5 ft 11 in) tall. Therefore, a measure of weight in relation to height has to be used to assess whether or not people are overweight or obese. A relatively simple calculation – first used by the Belgian statistician Adolphe Quételet (1796–1874) and originally called the Quételet index, but now called the **body mass index (BMI)** – is used to indicate whether an adult is a healthy weight for their height. To calculate your BMI, divide your weight (in kg) by your height (in m) squared.

$$BMI = \frac{weight\,(kg)}{height\,(m)^2} = \frac{weight\,(kg)}{height\,(m) \times height\,(m)}$$

From this equation, you can infer that the units of the BMI are kilograms per metre squared, written in scientific convention as $kg\,m^{-2}$ (and elsewhere as kg/m^2). However, in practice, BMIs are often not given any units.

There is not one perfect weight for every height, but a range that allows for people's different builds. As will be discussed shortly, this BMI formula is not suitable for children, nor (for obvious reasons) pregnant women and people with certain medical conditions. With these caveats in mind, the WHO recommends the use of the BMI as a guide for the classification of overweight and obese (Table 1.1). According to WHO guidelines issued in 2000, people with a BMI of less than 18.5 are considered underweight. Those with a BMI of 25 and up to 29.9 are defined as overweight and people with a BMI of 30 and up to 39.9 are defined as obese, or **clinically obese**, meaning that this is considered to be in itself a disease condition. People with a BMI of 40 and over are described as **morbidly obese**; morbid means diseased and these people will be exhibiting **comorbidities**, i.e. associated disease and disability.

Table 1.1 Body mass indexes and the associated WHO classifications.

Body mass index/$kg\,m^{-2}$	Classification
less than 18.5	underweight
18.5–24.9	desirable or healthy range
25–29.9	overweight
30–34.9	clinically obese (class I)
35–39.9	clinically obese (class II)
40 and over	morbidly obese (class III)

It is important to stress that these categorisations are arbitrary and that thresholds have been changed from time to time. However, this WHO classification system is widely used. Nevertheless, some studies choose to use different categories, as you will start to see in Section 1.4.1.

◆ What is the BMI of a woman who is 1.7 m tall and who weighs 88 kg?

◆ $\dfrac{88}{1.7^2} = \dfrac{88}{1.7 \times 1.7} = \dfrac{88}{2.89} = 30.4$

This value means that she would be in the obese range for her BMI. However, Figure 1.8 shows a young woman who has a BMI that classifies her as obese, yet she does not appear to be carrying too much fat. People who have developed above-average musculature will be of above-average weight for their height. Taken out of context, their BMIs are misleading. Nevertheless, such people are a minority, and when studying the health of large populations it is convenient to be able to obtain a universally recognised index by the simple method of measuring a person's height and weight. This WHO-endorsed system of measurement and classification is very useful as a basis for comparative and evaluative studies, such as those shown in Figures 1.2 and 1.3. However, care needs to be exercised in interpreting results because the relationship between BMI, **adiposity** (fatness) and the risk of ill health is not identical across all ethnic groups (see Section 1.6.3).

1.3.2 BMI in children

There is currently (2008) no universally accepted system for the classification of childhood obesity, although several-BMI based systems are available, including that of the WHO and the International Obesity Taskforce. The nub of the problem is that the relationship between the child's weight, height and amount of fat is not consistent over time. For example, water contributes around 65% of an adult's body weight. When a baby is born, water constitutes a much greater percentage of its weight (75–80%). There is currently insufficient information to know whether the reduction in water content with age occurs in a gradual way or not. Another difference is that the bones of a baby are much softer than those of an adult, containing a lower percentage of the minerals calcium and magnesium, and so are relatively less dense. So changes in weight as the child grows may reflect changes in body composition that are not related to fat deposition. It is often clear that, especially in boys, an increase in weight is the result of an increase in muscle rather than fat (see Chapter 6). Despite these difficulties, the UK Government in its white paper *Choosing Health?* (House of Commons Health Committee, 2004) pledged to reintroduce the school medical in England: the annual weighing and measuring of children in primary schools. This was opposed by the UK's Royal College of Nursing and other organisations. They argue that these measurements may embarrass and stigmatise young people. Furthermore, there has been debate around whether, once the measurements have been made, parents should be told if their child is considered to be obese – and the extent to which the state should intervene in these cases.

1.3.3 Measuring body fat

While recorded BMIs based on the relationship between weight and height measurements are undoubtedly a useful tool for population comparisons, other methods are favoured for research and for clinical settings. The question that needs to be answered in individual cases is: how much fat is this person carrying?

Methods for estimating the amount of fat in a person's body have become increasingly sophisticated. However, it has been shown that some very simple techniques work well too. The ratio of waist-to-hip circumference (waist–hip ratio or WHR) is a good approximate index of visceral fat as distinct from the subcutaneous abdominal adipose tissue. An even more convenient measure

Figure 1.8 Eva Massey, aged 26, has a BMI of 30.4. Eva hopes to compete in the Olympics as a shot putter.

Density is a measure of the mass per volume of a given substance or material.

is waist circumference, which correlates closely to BMI and WHR. Indeed, increases in waist circumference also correlate to increased risk for various chronic diseases (WHO, 2003).

For individuals, waist measurement may be a more accurate indicator of health risk than BMI and is easier to obtain. This is the basis of the UK's National Obesity Forum's campaign 'Waist Watch Action' (Figure 1.9a). They advise that measurement should be made at the level of the navel and that for women 81–88 cm (32–35 in) and for men 94–102 cm (37–40 in) indicate overweight. Values greater than 88 cm and 102 cm, respectively, indicate that the person is obese.

In a clinical setting, measuring the waist circumference of an obese person often presents a practical problem.

◆ Looking at Figure 1.9b, can you identify this practical problem?

◆ The health professional may not able to take the measurement on their own. A second person may be needed to ensure that the tape is in the correct position across the back. There may also be concerns if the health professional has to lean around someone who is partially clothed and not the same sex, which may be alleviated by the presence of a chaperon.

Figure 1.9 (a) Waist Watch Action, a campaign launched by the National Obesity Forum, encourages people to check their waist measurement. (b) Measuring the waist circumference of an obese person.

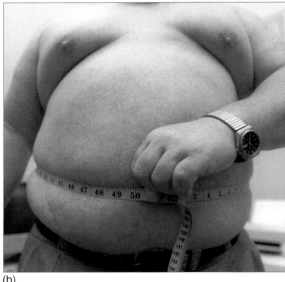

(a)   (b)

Another relatively straightforward method for estimating adiposity is the skinfold calliper (Figure 1.10). Skinfold thickness is measured at a number of specific sites on the body; from these measurements, the percentage of the body mass that is fat (adipose tissue) can be calculated. This value is then compared with the classifications shown in Figure 1.11.

◆ Compare the healthy ranges of body fat for men and women. In what way are they different?

◆ Women are considered to be healthy while carrying a greater percentage of body fat than men.

◆ Why do you suppose this to be a valid judgement? (Hint: Think about the differences in fat depot distribution that characterise the different genders.)

◆ Women have breasts that are composed of adipose tissue and also deposition of fat around the thighs and buttocks. These are responsible for their extra body fat content, but are not unhealthy features.

The use of skinfold callipers requires careful training as small measuring errors can lead to large errors in the final calculation. In the hands of a skilled practitioner, these callipers can provide a useful measure of progress in weight loss programmes, although the suggestion that they might be used should be approached with sensitivity as some people can find their use distressing.

◆ Why would the skinfold measure be a particularly useful measure for monitoring the progress of an obese person on a weight loss programme?

◆ The obese person needs to reduce their percentage of body fat. Although programmes designed to do this are often referred to as weight loss programmes, it is reduction in percentage of body fat that is important when seeking health gains. Also, it can be more convenient for a health professional to take a skinfold measurement than to take the waist circumference measurement.

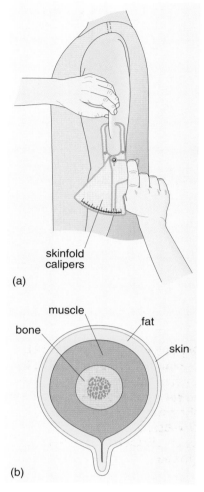

(a)

(b)

Figure 1.10 Skinfold measurement using a skinfold calliper. (a) Measurement being made on the upper arm. (b) Cross-section through the upper arm showing how the skin is pinched so that a double thickness of fat is measured.

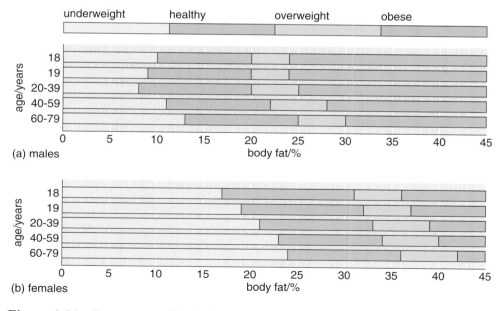

(a) males

(b) females

Figure 1.11 Percentage of body fat and the associated classifications for adult men and women of different ages.

Increasingly, machines that can scan the whole body are being used to assess body fat content (Figure 1.12). They make use of imaging techniques requiring X-rays or magnetic fields. These scanners are expensive to operate, but their use can sometimes be justified because the cost of ignoring obesity may be even greater given its association with ill health – a topic that will be explored in the next section.

MRI scanners use a strong magnetic field and radio waves to affect the hydrogen atoms in water molecules so that they give off signals, which the scanner then detects. MRI can produce detailed images of body tissues because different types of tissue have different water contents.

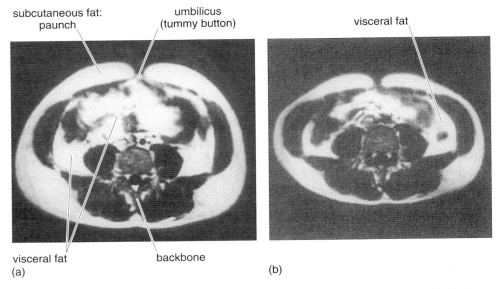

Figure 1.12 Magnetic resonance imaging (MRI) scan of an abdomen (a) before and (b) after 10% weight loss.

1.3.4 Summary

Epidemiological studies make broad comparisons between different groups (populations) of people, using BMI to assess the prevalence of overweight and obesity. The WHO categories are widely (but not universally) used. BMI relates weight and height to give an indication of adiposity, but other body measurements, such as waist circumference, WHR and skinfold thickness, are better methods for assessing an individual's adiposity. The best estimate of adiposity is gained using scanning techniques, but the cost cannot usually be justified.

1.4 The relationships between obesity and ill health

1.4.1 Does being overweight increase the risk of premature death?

If becoming obese were like acquiring wrinkles, merely a matter of personal regret and vanity, there would still be an industry offering prophylactic (preventative) and obesity-reversing measures – pills and potions and lifestyle advice – but, as with wrinkles, governments and public health services would not be interested. The reason that obesity interests these external agencies is that the comorbidities are a major financial burden to the community.

Indeed, it was actuaries (persons who calculate the likelihood of disease and premature death for insurance companies) who in 1959 produced figures showing that life expectancy was lowered for those who were obese. A more recent study has demonstrated that this is also true for overweight people (Adams et al., 2006). We will use this study to highlight some of the difficulties in obtaining information about populations – and some of the methods used to overcome these difficulties.

The study followed 186 000 men and women aged between 50 and 71 (when recruited) over a 10-year period. It is called a *prospective* or *cohort* study (Figure 1.13) because once the group of people had been recruited to the study, information was collected from them on several occasions over a considerable time period. The information was gathered from self-reported responses to a questionnaire. Inevitably, there will be a certain level of inaccuracy; in particular, it is known that overweight and obese individuals are inclined to represent themselves as being taller and less heavy than they actually are. The overall study involved more than half a million people, but anyone who had ever smoked was excluded from this part of the analysis because smoking is a risk factor for a number of diseases that lead to premature death. In other words, individuals who smoke are at increased risk of dying at an earlier age than non-smoking individuals.

Figure 1.13   Cohort studies are named after cohorts – groups of Roman legionnaires that were formed and continued without new members until the last of them had died.

There are a number of other factors that affect an individual's risk of dying. Those that were taken into consideration in this study were whether or not the individual already had a chronic disease (a disease that has been established for a long time), their ethnic group and their age and gender. In the 10 years from the beginning of the study, twice as many men as women died, so data for men and women were shown separately. For the other factors, it was necessary to perform mathematical adjustments to standardise the data (see Box 1.2).

Box 1.2   Standardising data

In an ideal experimental situation, the scientist compares two groups which are identical in every respect except for one *variable*, which is the factor that interests the scientist. So if interested in the effect that being overweight has on life expectancy, two groups would be selected, one overweight, one healthy weight, but made up of individuals who were in every other respect identical: identical lifestyle, occupations, wealth, marital status, age, ethnicity, etc.

As you can see, this is not going to be possible in the kinds of epidemiological studies we've been looking at. Instead, to ensure that these factors do not interfere with the investigation into the one variable of interest, it is necessary to perform mathematical adjustments to standardise the data. For example, other things being equal, individuals in their 70s have a greater chance of dying than those who are in their 50s. If there were a great number of people of healthy weight in their 70s in the study, then the number of deaths recorded from this group might be considerably greater than the number of deaths recorded from an obese group.

Suppose there were 1000 people of healthy weight in the 70–75 year age group, of whom 10 died. Their **mortality rate** (deaths per 1000 of the population) would be 10. If 10 people who were obese and were 70–75 years old also died in the same period of time, it might appear superficially as though being obese makes no difference to your risk of death. However, if you were then told that the population of obese 70- to 75-year-olds was 200, the conclusions you would draw would be different. Their death rate is 10 in 200, giving a mortality rate of 50 per 1000 of the population. This is 5 times more than for the healthy weight group.

An adjustment, called **age standardisation**, can take account of differences in population age structures between groups whose data you want to compare. The method involves taking a 'standard population' or group (such as all the individuals who had had a healthy BMI at age 50) and using this population's age structure as the basis for adjusting the data from the other groups to what it would be if they had the same age distribution as the 'standard population'.

Figure 1.14 shows the relative risks of death in relation to the BMI recorded at 50 years old, among men and women who had never smoked. In order to understand Figure 1.14, it is necessary to explore how risk is defined and measured. Risk is the probability of something unpleasant happening. In this study, individuals with a BMI of 23.5 to 24.9 were taken as the reference group – that is, the group with whom other groups were compared. They were chosen as being a group with a healthy weight for their height. To make the calculations easy, let us suppose that in this non-smoking reference group the risk of dying in any one year is 2 per 1000 in the population. And let us suppose that in a non-smoking group with a BMI of 30 the risk of dying in any one year is 4 per 1000. The obese group has double the death rate of the reference group. This is called

Note that this BMI range is not one of the WHO categories.

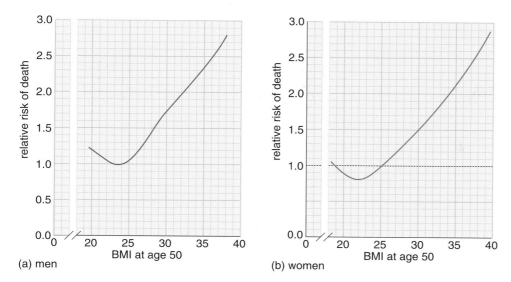

Figure 1.14 The relative risks of death in relation to BMI among (a) men who had never smoked and (b) women who had never smoked.

the **relative risk** (the higher rate divided by the lower rate). In order to present this information graphically, the reference group's risk is assigned a value of 1.0, so the group with the BMI of 30 will then have a relative risk value of 2.0.

◆ From Figure 1.14, determine the relative risk of death for men who have never smoked with a BMI of 38.

◆ The relative risk of death for men who have never smoked with a BMI of 38 is 2.8: nearly 3 times as great as for the men with a healthy BMI.

◆ What do you note about the relative risk of death for women who have never smoked and have a BMI of 22?

◆ These women have a relative risk of 0.8. This means they have less chance of dying than the women in the group chosen to be the reference group.

Adams and his colleagues concluded from the study of the whole cohort that if a person is overweight in their middle years, their risk of death is increased. This conclusion is shown particularly clearly in Figure 1.14 using the results from the non-smoking cohort, but they found their conclusions held for all individuals, regardless of gender, age or ethnic group – all of which are, of course, *non-modifiable* risk factors.

Not all studies of the risks associated with being overweight have come to the same conclusions. One study (Flegal et al., 2007) was greeted with newspaper headlines of 'Doctors say being fat is good for you'. This was not quite what the authors of the study were saying! Also, they had studied a different group or cohort from Adams and his colleagues – and in a different way. For example, Flegal and colleagues had individuals from age 25 upward, and they had identified groups from health records and looked back (rather than recruiting the groups and following them). (This is a *retrospective* study; both prospective and retrospective studies are known as **longitudinal studies**.) Their study showed that being overweight increased a person's relative risk of dying from diabetes or kidney diseases, although overall the risk of premature death was less than for the healthy weight group – which they had chosen in line with WHO guidelines (i.e. BMI of 18.5–25). One suggestion that they made for this finding was that if an overweight person becomes ill and starts to lose weight, they have more reserve to fall back upon than a person of healthy weight.

Although there may be differences in opinion on the exact changes in risk associated with being overweight (i.e. BMI of 25–29.9), all studies show that obesity – especially as it becomes more extreme – is associated with major health risks.

For younger age groups, the WHO warns of far more dangerous consequences based on their assessment of a number of other studies. For morbidly obese people in the 25–35 age group, they suggest there is a 12-fold increased risk of mortality when compared with individuals of healthy weight.

Although obesity is classified as a disease, and therefore it can and does appear on death certificates as the cause of death, the majority of obese people die from other diseases. Most of these are diseases that have developed as a consequence of the excess body fat.

1.4.2   The diseases associated with obesity

The WHO's assessment of the major risks of ill health associated with obesity – and the relative risks of an obese person developing them – are given in Table 1.2 (WHO, 2002). For the moment, do not be concerned if you are not familiar with all the terms; they will be described and discussed in Chapter 7.

Table 1.2 Relative risk of health problems associated with obesity in adults.

Greatly increased (relative risk much greater than 3-fold)	Moderately increased (relative risk between 2- and 3-fold)	Slightly increased (relative risk up to 2-fold)
type 2 diabetes	cardiovascular diseases	cancer
gall-bladder disease	hypertension	impaired fertility
dyslipidaemia (abnormal concentration of fats in the blood)	osteoarthritis (in knees)	lower back pain
insulin resistance		risk of anaesthesia complications
breathlessness		fetal defects associated with maternal obesity
sleep apnoea (disturbance of breathing while asleep)		

All relative risk values are approximate. They are difficult to estimate because of the variability between individuals and the fact that these individuals may have other conditions or lifestyle factors, such as drinking or smoking, that predispose to ill health.

It is noteworthy that Table 1.2 makes no mention of any mental health problems, such as depression. Opinions about the relationship between obesity and depression have changed over the years. In many instances, therapists suggest that obesity develops as a result of overeating in response to adverse life events, but there is also evidence of some obese people becoming clinically depressed as a consequence of their predicament. More will be said about this in Chapters 7 and 9.

When considering the health of populations, obesity has its greatest impact on health through its effect on diabetes. It has been calculated, on the basis of a number of studies, that overweight and obesity accounts for almost 65% of diabetes cases in men and over 75% of cases in women (Seidell, 2005). Furthermore, 60% of people with diabetes, as compared with 35% of the general population, die from cardiovascular diseases (Watkins, 2003).

◆ Is a person who is obese and has developed diabetes any more likely to die from cardiovascular causes than someone who has diabetes but is not obese?

◆ Yes, they are, because being obese places them at risk of developing cardiovascular diseases independently of the risk associated with having diabetes (see Table 1.2).

It is of particular concern that type 2 diabetes, which used to be called maturity-onset diabetes, is now being seen in young children. It is estimated that over 20 000 children in the European Union (EU) countries between 5 and 17 years of age have diabetes and 400 000 have impaired glucose tolerance, a condition that usually leads to developing diabetes (Lobstein and Jackson Leach, 2007). Figure 1.4 shows how the number of overweight children has been increasing in EU countries in recent decades, and the fear is that this will result in earlier onset and increased prevalence of diabetes. Inevitably, there will be other comorbidities and complications, particularly relating to the cardiovascular system, and many premature deaths.

Impaired glucose tolerance will be explained in Chapter 7.

Ill health associated with obesity and its consequences represents considerable costs to individuals, their families, governments and society. In the next section, we explore the economic cost of obesity.

1.4.3 Summary

There is evidence that if a person is overweight in their middle years, their risk of death is increased. There are, however, many other lifestyle as well as non-modifiable factors, such as gender, inherited diseases and ethnicity, that can decrease life expectancy too. All of these factors must be considered and adjustments made to standardise the data to allow a fair account of the results of epidemiological studies to be presented. Obesity shortens life expectancy, but is associated with a number of diseases that do not kill quickly, instead reducing quality of life over an extended period.

1.5 The economic cost of obesity

It is estimated that overweight and obesity accounts for 5% of total costs of the UK's National Health Service (NHS). This is comparable with estimates of 4% for the Dutch health care system and better than the estimated 8% for the USA. However, France does rather better, with obesity costing only 2% of their health care budget (Allender and Rayner, 2007). A glance back at Figure 1.1 should give you an indication of the reasons for these different spending patterns.

The economic costs of obesity are estimated by making assumptions about the extent to which obesity has contributed to various other diseases. Table 1.3 gives some insight into how these costs are calculated, using records of costs collected by the NHS in the UK. The second column of Table 1.3 shows the costs that relate to diseases for which being overweight and obesity are risk factors. However, there will be patients who are of healthy weight being treated for these diseases, so an adjustment must be made by applying a correction factor to the costs, shown in the third column. This correction factor is based on the proportion of the population with the disease who probably have the disease *because of their weight*, combined with an adjustment that takes into account each individual's life expectancy. The rationale for this is that if life expectancy is lowered for a person with a particular BMI at a particular age, then the length of time they are treated by the NHS is also reduced, and hence cost is also reduced.

Table 1.3 Estimates of National Health Service (NHS) costs attributable to diseases related to overweight and obesity in 2002 in the UK.

Cause	Cost/ £ millions	Correction factor (proportion of cost attributed to overweight and obesity)/%	Estimated cost because of overweight and obesity/£ millions
heart disease	2287	34	778
stroke	2892	34	983
breast cancer	240	12	29
colon/rectum cancer	383	16	61
hypertensive diseases	994	58	576
uterine cancer	85	49	41
osteoarthritis	1090	21	229
type 2 diabetes	675	79	533
total	8645		3231

◆ In terms of cost, through which disease did being overweight and obese place the greatest burden on the NHS in 2002?

◆ Stroke, which, as a consequence of overweight and obesity, cost the NHS £983 million.

◆ Which disease ascribes the highest proportion of its cost to being overweight and obese, and what was the attributable cost to the NHS in 2002?

◆ 79% of the cost of treating diabetes is attributed to overweight and obesity. The cost to the NHS was £533 million.

(Recall from Section 1.4.2 that another study suggests that over 65% of cases of diabetes in women and 75% in men were attributable to overweight and obesity. It is inevitable that different studies will come up with different estimates – but these values are similar to that in Table 1.3.)

So although obesity gives a greater risk of developing diabetes than cardiovascular diseases (Table 1.2), it is the cost of treating these cardiovascular diseases – such as stroke and heart disease – that places the greater burden on the NHS.

Table 1.3 shows a total cost to the NHS in 2002 of £3.2 billion for treating disease that is directly attributable to overweight and obesity. Allender and Rayner went on to suggest that over 66 000 deaths in 2003/2004 could have been avoided if the population could have achieved a healthy BMI of around 21. (Note that they did not choose a range of values to describe a healthy BMI, although the value chosen does fall within the WHO guidelines.)

Now look at Table 1.4, which also estimates the costs of overweight and obesity to the NHS in 2002 (House of Commons Health Committee, 2004). It looks at costs in relation to categories of spending, such as the cost of consultations with GPs (general practitioners, who are local doctors), rather than categories of disease.

Recall from Section 1.1 that a billion is one thousand million.

Table 1.4 The estimated cost of overweight and obesity to this NHS in 2002. GP, general practitioner (local doctor).

	Category of spending	Estimated cost because of overweight and obesity/£ millions
treating the consequences	GP consultations	180–210
	hospital admissions	420–500
	day cases	20–30
	outpatient attendances	120–180
	prescriptions	1150–1250
	total	1890–2150
indirect costs	lost earnings due to attributable mortality	2100–2300
	lost earnings due to attributable sickness	2600–2900
	total	4700–5200

◈ Compare Table 1.3 with Table 1.4. How do their estimates differ of total costs to the NHS in 2002 for treating disease that is the consequence of overweight and obesity?

◆ In Table 1.4, the costs attributable to the NHS from treating disease related to overweight and obesity are estimated at around £2 billion (£1890–2150 million). This is about two-thirds of the estimate of £3.2 billion shown in Table 1.3.

So why is there this big difference between the two tables? Both are from reputable sources, but Table 1.3 is based on more recent information about the actual level of obesity in the UK in 2002, as well as clearer ideas on the method of attributing risk. Table 1.4, published in 2004, assumes a level of obesity of 20%, whereas by 2007, with more information available, the official estimates were that the level of obesity in 2002 was 22% for women and 23% for men. In general, it is this kind of uncertainty associated with the difficulties in obtaining up-to-date and reliable data that makes it hard for policy makers to decide where best to focus resources to attempt to halt the increase in levels of overweight and obesity and to ameliorate the effects of the condition. This will be discussed further in Chapter 10.

Table 1.4 draws attention to other costs attributable to overweight and obesity. These are termed indirect costs. If individuals are not earning because of sickness (or death) then they are not paying taxes, so there is a notional loss to the government and to society. If there is a need to pay benefits such as sick pay or income support to bereaved families from government resources, then this further increases the indirect cost of the condition. Furthermore, obesity kills slowly; most obese people live to pensionable age, yet may have been unable to work and contribute to a pension scheme. They often require above-average levels of social care as they age because of difficulties such as immobility. There is more about these costs to individuals in Chapter 7.

The costs of different items that are needed to provide direct health care are also shown in Table 1.4.

◆ Which are the two most costly items in treating diseases that are a consequence of overweight and obesity?

◆ The most costly item is providing medication (prescriptions). The second most costly item is hospital treatment, but this is less than half the amount spent on prescription drugs.

The consequences of obesity, and associated attitudes, will be addressed more fully in Chapter 7, where we will also consider the social and psychological costs to individuals.

The number of people affected by obesity and the costs involved indicate a major public health problem that has gone beyond the point where it can be tackled by individuals alone. An important consideration for government policy makers is whether some groups of the population are more at risk than others. The groups and their risks is the topic of the next section.

1.5.1 Summary

The NHS costs for treating obesity are calculated by including the costs of treating diseases associated with overweight and obesity. Obesity is most strongly associated with diabetes, but the costs associated with treating obesity-related cardiovascular diseases are the greatest financial burden on the NHS. The estimates of costs incurred require certain assumptions to be made, which can dramatically affect the final estimates. This kind of uncertainty makes it hard for policy makers to decide on priorities for spending and the best strategies for challenging the obesity epidemic.

1.6 Who is most likely to become obese?

In this section, we once again draw on epidemiological evidence to highlight associations, but we will explain in Chapter 6 *why* particular groups are more at risk than others.

1.6.1   The effect of gender

Most studies have shown the prevalence of obesity to be higher in women than in men. However, the differences are not large and they appear to be narrowing – certainly in the UK (Figure 1.15). In Chapter 6, you will learn that there are perhaps rather more factors that predispose women to gain more weight than men. Vanity may have helped to narrow the gap. There is certainly much pressure from the media and advertising for women to maintain a slim figure. The 'Tackling Obesities' project mentioned in Section 1.1 predicts that only the most privileged women (Social Class I) in the UK will resist the trend towards steeply increased levels of obesity. The prediction is that, by 2050, 15% of Social Class I women will be obese compared with 62% for the least privileged women (Social Class V). The predicted levels for men are 52% for Social Class I men compared with 60% for Social Class V.

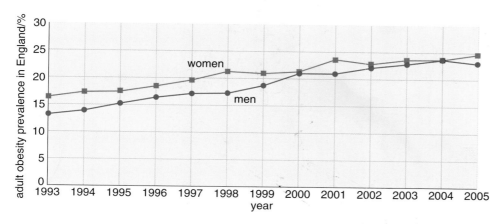

Figure 1.15   Prevalence of obesity in adult men and women in England, 1993–2005.

◆ For England, in which year did the prevalence of obesity in men first reach the same level as the prevalence of obesity in women?

◆ In 2000.

1.6.2   The effect of age

The prevalence of obesity increases with age up until around 65 years old. This is illustrated by Figure 1.16, which shows data collected in England in 2002.

◆ First refer back to Figure 1.15. Was the overall prevalence higher for men or for women in that year?

◆ There was little difference in prevalence.

◆ Describe the differences in prevalence for men and for women shown in Figure 1.16.

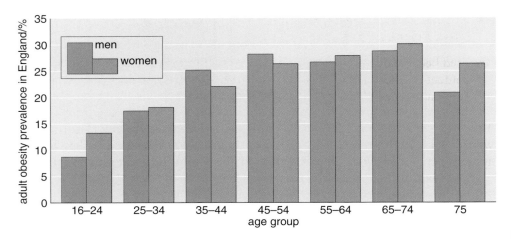

Figure 1.16 Prevalence of obesity in England in men and women in 2002 by age group.

◆ For most age groups, the prevalence of obesity is higher for women, but between the ages of 35 and 55, the prevalence is higher in men.

◆ What could be the reason for this?

◆ One possible reason is that more young men than young women take part in active sports in early life and often quench their thirst at the end of a match with beer, which can add significantly to their energy intake. When they give up playing, they may continue to watch the sport and naturally also consume the beer. Later in life, women go through the menopause. Hormonal changes associated with the menopause may be responsible for the higher prevalence of obesity in women than in men beyond the age of 55.

Figure 1.16 is based on the BMI definition of obesity. However, there is also a tendency for the *percentage* of body fat to increase with age. So someone who has maintained a fixed weight as they age – and therefore a static BMI – can, nevertheless, become obese where obesity is defined by the percentage of body fat, as shown in Figure 1.17.

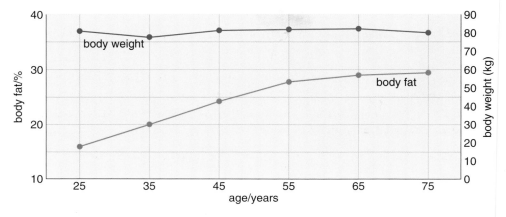

Figure 1.17 Age-related increase in body fat for Fred, a healthy man with a constant BMI.

◆ At what age did Fred become overweight, and at what age was he obese? (Hint: Refer back to Figure 1.11.)

◆ Fred became overweight sometime around 35 years old (when his body fat content reached 20%) and became obese by 57 years old (when his body fat content reached 28%).

This suggests that, as people age, they should aim for a slight decrease in weight, whereas the observation is that many people feel they can relax a bit and that 'middle-aged spread' is inevitable. There is more on the effects of age on obesity in Chapter 6.

1.6.3 The effect of ethnicity

The effect of ethnicity is evident when people of different ethnic backgrounds are living in the same country (Figure 1.18).

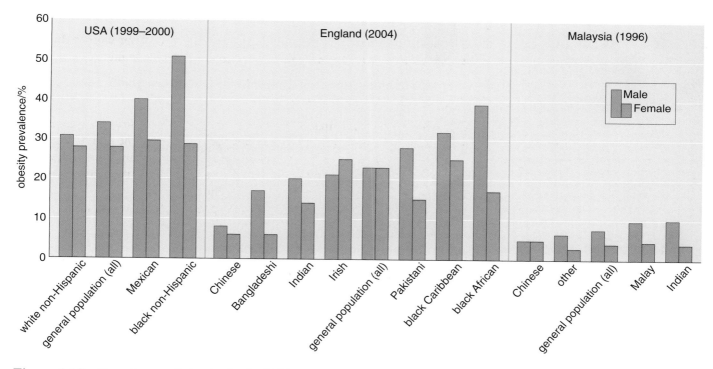

Figure 1.18 Prevalence of obesity in the USA, England and Malaysia by gender and ethnic group.

◆ Study Figure 1.18 and note the obesity prevalence for Indian populations who were living in England in 2004 and those who were living in Malaysia in 1996. In relation to these two populations, what is the most striking feature shown in these bar charts?

◆ Indians have a much higher obesity prevalence when residing in England.

◆ Apart from their genetic make-up, in what other way might ethnic groups differ that could contribute to the differences shown in Figure 1.18?

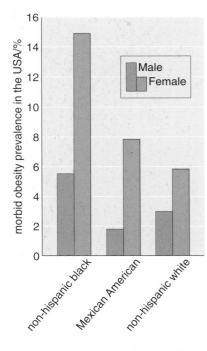

Figure 1.19 Prevalence of morbid obesity in US adults, 2003–2004.

◆ These debates continued until in 1914 an Act of Parliament brought in the provision of free school meals. It was about 90 years later that Parliament was again debating school meals, but this time in relation to unhealthy eating and obesity.

◆ The groups might differ in their socioeconomic status, in their culture, in their attitudes towards body size, in the types of food they cook and in the way food is used in social settings.

So although everyone living in the USA, in England or in Malaysia might experience a broadly similar environment within that country, the choices people make within that environment may differ.

The rates of morbid obesity are of special concern. Figure 1.19 shows rates for morbid obesity in men and women from three different ethnic groupings in the USA.

◇ What is meant by morbid obesity?

◆ Morbid obesity is any BMI of 40 and over (see Table 1.1) and is typically associated with other diseases.

◇ What is the prevalence of morbid obesity among black American women in the USA?

◆ Over 14% of black American women in the USA are morbidly obese.

Black American women also have higher levels of overweight and obesity than the general population. However, there is evidence that the relationship between obesity and its comorbidities may vary between different ethnic groups (Neovius et al., 2004). Thus for the same level of obesity the risk of acquiring another specific disease can differ between ethnic groups. A number of studies show that Asian populations, including Indian, Malay, Chinese, Indonesian and Thai peoples, have a greater risk of comorbidities at a lower BMI than Caucasians. This is why studies, such as the one reported by Adams and colleagues in Section 1.4.1, make adjustments for ethnicity.

We will return to a consideration of how ethnic differences might arise in Chapter 6.

1.6.4 The effect of socioeconomic factors

Historically, to be fat was a sign of wealth. Only the rich could afford an abundant diet. Even in more recent times, at the beginning of the 20th century, the reigning British monarch, Edward VII (Figure 1.20), reportedly had a 46-inch (117-cm) waist at a time when the plight of undernourished children was the subject of considerable parliamentary debate.

Being fat is still a mark of status in some areas of some countries (e.g. parts of the Middle East, Mauritania and the Indian subcontinent; see Vignette 1.2). A plump bride has often been particularly desirable and there are still 'schools' for fattening young girls in Mauritania. While obesity can develop as a response to feeling miserable, it can also develop as a response to feeling content. Typically, people of any age increase their BMI within 2 years of marrying or forming a similar settled relationship.

Vignette 1.2   Introducing Nazneen

Nazneen Iqbal was brought up in Bangladesh but moved to Yorkshire when she was 20 years old to marry Tariq. She had a BMI of 27 when she arrived in the UK. Nazneen learned from her mother that cooking and providing food was the sign of a good wife. Tariq was delighted with her traditional cooking and the fact that she used generous amounts of ghee (clarified butter). Tariq and Nazneen are oblivious to the fact that their weights are gradually increasing.

You will read more about Nazneen in Chapter 7.

Figure 1.20   The British monarch King Edward VII (1841–1910; ruled 1901–1910). Do you think he was obese? A waist measurement above 40 inches (102 cm) indicates obesity, so at 46 inches (117 cm), he is above this threshold.

It is still the case that in many less developed countries it is the wealthy who are obese, as only they have the economic means to overindulge. In more developed countries, by contrast, those with the lowest incomes are more likely to be obese. This can be neatly illustrated using data from Brazil (Figure 1.21), which has, over time, become a more developed country.

◆ Look at Figure 1.21. Did the wealthiest group of women in south-east Brazil have the highest prevalence of obesity in 1975?

◆ No, the wealthiest group of women had the second-highest prevalence of obesity, at just over 9%; it was the next wealthiest group who had the highest prevalence, at just over 10%.

◆ In 1975, which group had the lowest prevalence of obesity?

◆ The lowest level of obesity was seen in the least wealthy group of women.

This illustrates that the wealthy are more likely to be obese than those with the lowest incomes in less developed countries.

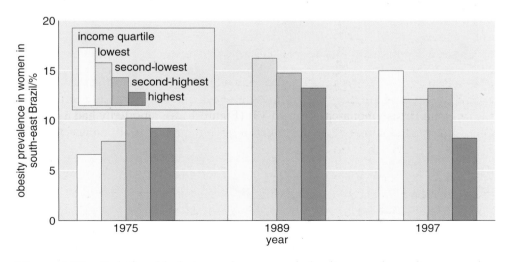

Figure 1.21   Relationship between income and obesity prevalence in women in south-east Brazil, 1975–1997.

◆ How had the situation changed by 1997?

◆ By 1997, the least wealthy women were the most obese (prevalence was 15%), whereas in the wealthiest group the prevalence of obesity was the lowest (at just over 8%).

In the two decades after 1975, the population in south-east Brazil had become more prosperous.

Education, income level and class are all associated with obesity prevalence, and it can be difficult to separate out the individual effects of these factors. Within most developed societies, it is the poorly educated and poorly paid women in manual occupations who are most likely to be obese. The children of working women also tend to be obese. While there are many campaigns to inform the public about healthy eating and the importance of being physically active, it appears that so much information leads to confusion (see Vignette 1.3).

Vignette 1.3   More about Charlie

Charlie's parents are not well off. His dad, Dave, works in a warehouse and his mum, Maureen, is a cleaner at the local dairy. Charlie is their only and much-loved child. Maureen really regrets that she can't be there when he comes home from school and encourages him to invite friends back so that he has company. She knows how hungry young lads can get, so she always leaves them snacks; she leaves savoury snacks, crisps (potato chips) and nuts because she thinks that is better for them than sweet, sugary snacks.

When she gets in from work, she gets a meal ready for the family, and she and her husband enjoy hearing about Charlie's day. She is often tired and uses ready meals or buys a 'takeaway' on her way home, but she is a good cook and they always have a proper Sunday lunch with roast potatoes and an old-fashioned pudding, such as treacle tart.

For some individuals and their families, the relationship between poverty and obesity can develop into a downward spiral. The overweight child can experience negative attitudes at school and have a diminished educational experience. The obese individual often experiences discrimination, making it difficult to get a job at all. Health care may be denied or not accessible for financial reasons in some countries. These issues will be discussed in later chapters; in the next section, we begin our examination of the causes of obesity.

1.7   Why have so many people become overweight and obese?

The simple answer to this question was given in the quotation at the start of this chapter. People have poor eating habits and they get fatter when they eat more than they need for their daily activities. But how do the people who stay slim and have low adiposity know how much they need to eat each day? The answer is that they don't 'know' how much to eat any more than they 'know' when to take their next breath or produce their next heart beat. The body has a remarkable ability to regulate its activities without any conscious input being made, as you will learn in the next chapter. But excess food intake leads to excess body weight, which leads to a variety of diseases. So why are humans so prone to eating more than they need?

A speculative answer to this question is provided by Prentice and Jebb (2003). They suggest that the critical point to consider is the environment in which our human ancestors lived. Genetic and fossil evidence suggests that modern humans, *Homo sapiens*, evolved in Africa around 120 000–200 000 years ago from an ancient *Homo* species, which had a hunter–gatherer lifestyle. Since then, for most of the time, humans have also been hunter–gatherers, until about 12 000 years ago, when subsistence farming became established. Neither the hunter–gatherer nor the subsistence farmer had access to the kinds of food that are available in many developed societies. Even today, some groups of subsistence farmers exist by eating food that has, gram for gram, only a quarter of the energy value of a typical 'fast food' meal. Their diet has a low **energy density**.

◆ What do you suppose is meant by 'high energy density'?

◆ A food with a high energy density will make available far more energy per gram than food with a low energy density.

So it may be that humans are not good at regulating their food intake because it is only very recently that diets of high energy density have been widely available. In Chapter 5, you will assess evidence that humans are not good at recognising when they are eating a diet with a high energy density. Also, it seems that humans do not have the appropriate physiological and behavioural responses to such (potentially fattening) foods.

Another important difference is that modern humans living in developed societies do not take as much exercise as their ancestors. In other words, most humans are now living in an **obesogenic environment**: an environment in which food is both attractive and readily available, and in which technological and societal changes have reduced the impetuses for physical activity, thereby encouraging the development of obesity. In the next chapter, we start to assess how this has influenced the prevalence of obesity.

'Fast food' refers to modern convenience food that can be prepared, served and eaten quickly, and which is frequently high in fat, sugar and salt. Its origins are in the USA, but the archetypal meal of beefburger, fries and sugary carbonated drink is now available worldwide.

1.8   Summary of Chapter 1

1.1   Overweight and obesity are at epidemic levels and are predicted to continue to increase. The increasing prevalence in children is particularly worrying.

1.2   Obesity is recognised internationally as a disease and is defined as an excess accumulation of body fat that predisposes to ill health.

1.3   Fat is stored in adipose tissue. Adipose tissue is an endocrine organ; different depots release different hormones according to circumstances.

1.4   BMI is a useful measure of adiposity for epidemiological studies, but other methods are preferred for the assessment of individual adiposity.

1.5   Overweight and obesity are both associated with increased risk of developing comorbidities, particularly diabetes and cardiovascular diseases.

1.6   The costs of treating obesity and its consequences are considerable but are not easy to estimate accurately. Obese individuals are not killed rapidly by comorbidities, but tend to have increased disability and reduced life expectancy.

1.7 Individuals in some groups are at greater risk of becoming obese than others, but the reasons for this have not been explored in this chapter.

1.8 The obesogenic environment is very different from the environment that humans occupied for most of the past 200 000 years, when overcoming food shortages was often a pressing need.

Learning outcomes for Chapter 1

LO 1.1 Define and use, or recognise definitions and applications of, each of the terms printed in **bold** in the text.

LO 1.2 Explain how epidemiological evidence is used in the study of overweight and obesity.

LO 1.3 Describe and explain the uses and the limitations of different measures of obesity.

LO 1.4 Appreciate the contributions of different disciplines (economic, social, biological and medical) to understanding obesity.

LO 1.5 Demonstrate the ability to read a table and a graph and to calculate BMI.

Self-assessment questions for Chapter 1

Question 1.1 (LOs 1.1 and 1.2)

Within one month of the publication of the UK Government's *Tackling Obesities* report (discussed in Section 1.1), the suggestion that obesity-related diseases would cost an extra £45.5 billion per year by 2050 was being challenged by an academic from the University of Cambridge. Without knowing the specific grounds for this challenge, but based on your reading of Chapter 1, suggest how different academics could come to different conclusions on this topic.

Question 1.2 (LOs 1.3 and 1.5)

Sarah is an enthusiastic hockey player in her county team. She trains hard and hopes to make the England team. She is 1.55 m tall and weighs 77 kg. Kirsty is the same age as Sarah and has one toddler and a part-time job as a receptionist. She also weighs 77 kg and is 1.55 m tall. Calculate these young women's BMI. According to WHO guidelines, how would they be classified? Comment on the appropriateness or otherwise of this system of classification.

Question 1.3 (LO 1.4)

From your reading of Chapter 1, identify one example each of how (a) biology, (b) medicine, (c) economics and (d) social studies have contributed to your understanding of obesity.

Question 1.4 (LOs 1.4 and 1.5)

Examine Figure 1.2 and describe in general terms the difference in the data between genders for Mauritania. Explain the difference.

CHAPTER 2

ENERGY: INTAKE AND NEEDS

One of the few incontrovertible facts about obesity is that weight is only gained when energy intake exceeds energy needs for a prolonged period.

Jebb (2007)

2.1 Introduction

There is only one way in which humans can obtain usable energy, and that is from the food and drink they consume. Different foods provide different amounts of energy. (You have already met the idea of foods of high and low energy density in Section 1.7.) Most of the energy in food is released by processes in the body that also require oxygen to be available. The oxygen is needed for the chemical reactions that release the energy stored within the structures of particular types of food. These energy-releasing processes take place in every cell of your body and will be discussed in Section 2.6. Oxygen is obtained from the air when it is inhaled into the lungs and is carried by the bloodstream to all the body's cells. So oxygen is available all the time; stop breathing and you die. But food is different. You breathe continuously but you do not eat continuously. Yet the body requires a continuous supply of energy. In Chapter 4, you will find out how the body manages to keep its cells supplied with a continuous energy source despite a fluctuating intake of food.

Food has many attributes. At a practical level, it provides material to enable growth and renewal of body tissues as well as providing energy. At an emotional level, it often has pleasurable associations, providing a focus for social gatherings or a respite from boring occupations. In this chapter, the emphasis will be on understanding the chemical nature of food as it relates to providing the body with an energy source.

Energy is used for physical activity and a myriad of other processes that take place in the body, ranging from growth and repair of tissues to 'house-keeping' activities such as breathing, sleeping, thinking and digesting food. Underlying all such activities are chemical processes taking place in the body. **Metabolism** is the collective name given to these chemical processes.

Different individuals have different energy requirements. An individual is in a state of energy balance if the following equation applies to them:

total energy intake (TEI) = total energy expenditure (TEE)

When energy intake is greater than energy expenditure, the surplus will most frequently be stored as fat, and the individual is said to be in a state of **positive energy balance**. (An unfortunate term, as most people feel anything but positive about their weight gain.) Conversely, if energy intake is less than energy expenditure, the individual is in a state of **negative energy balance** and they will lose weight as fat stores in the body are mobilised to provide energy.

2.2   Energy intake

To measure your energy intake exactly is a difficult and tedious affair. Nor is it straightforward because foods (such as eggs, tomatoes, cakes, cheese and apples) consist of a number of components, different amounts of each component and not all of the components of food provide energy. For example, if you enjoy following a recipe to make a particular cake, you might not weigh out *exactly* the same amount of butter every time, and the eggs you use will be very unlikely to weigh *exactly* the same every time! Even apples from the same tree will not all have the same calorific (energy) value; as well as size differences, the proportions of sugars alter as the fruit ripens.

The components of food that are digested, absorbed and essential for use in bodily processes are known as **nutrients**. These are divided into:

- **macronutrients** (fats, proteins and carbohydrates): food groups that provide energy
- **micronutrients** (vitamins and minerals): food groups that are essential for body functioning, but which are required in small amounts and do not provide an energy source for the body
- water.

Macro- and micro- are derived from the Greek *macros* and *micros*, meaning large and small, respectively.

In order to lose weight, it is necessary to move into a state of negative energy balance, so it would seem obvious that it is necessary to reduce intake of the macronutrients, i.e. eat less fat, carbohydrate and protein. However, the macronutrients cannot be entirely omitted from the diet because, in addition to providing energy, they provide other essential materials, some of which cannot be stored but must be provided continuously. Therefore, it is important to maintain a balanced diet. A balanced diet contains six key nutrient groups (the macronutrients, micronutrients and water) in appropriate amounts for health. One difficulty with advising on a balanced diet is that 'appropriate amounts' depend on individual levels of activity and weight as well as age, gender and ethnicity. Although there are general guidelines advising on healthy, balanced diets issued by governments and the WHO, anyone with a serious weight problem needs to consult their doctor, who will enrol the support of a dietitian (a person with a professional training in nutrition and diet) to devise a personal plan for weight loss. Unfortunately in many countries, including the UK, this is not always possible as there are too few dietitians to meet the demand.

In the rest of this chapter, we briefly consider the guidelines offered for achieving a balanced diet and then explore in detail the chemical nature of the macronutrients so that you can understand how energy can be released from these substances. Finally, we see how the body uses up these resources in various activities, making it essential that we eat regularly.

2.3   What is a balanced diet?

Figure 2.1 shows the balance of foods in the diet recommended by the UK Food Standards Agency (FSA).

◆ Name the essential nutrients.

◆ Fats, carbohydrates, proteins, vitamins, minerals and water.

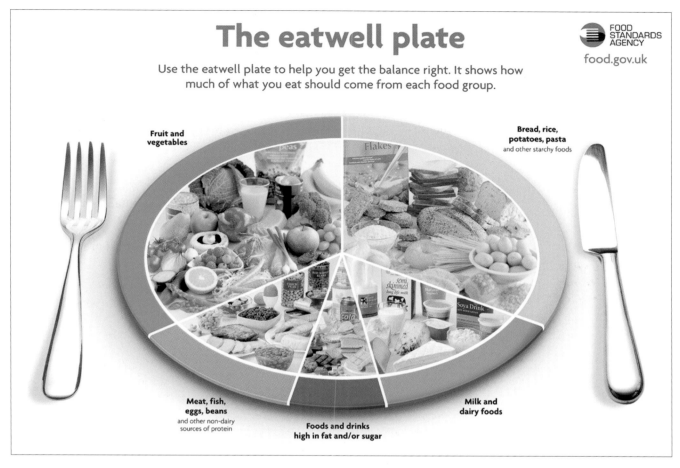

Figure 2.1 The 'eatwell plate': a guide to choosing a healthy, balanced diet.

◆ Does Figure 2.1 enable you to identify any specific nutrient that might be overrepresented in your diet?

◆ Figure 2.1 is a representation known as a pie chart. It enables you to see the relative proportions of each of the food categories that are likely to make up your diet. It is a fairly crude guide and does not allow you to identify any nutrient that might be overrepresented in your diet.

You might have noticed that the food categories used by the FSA do not equate to the nutrient groups. For example, fruits and vegetables are sources of carbohydrate but some of them also contain protein and fat (e.g. bananas, avocado and melon).

◆ What proportion of your diet should be composed of fruit and vegetables?

◆ Fruit and vegetables should make up a third of your daily intake.

The message in Figure 2.1 is simple and it emphasises balance rather than focusing on specific nutrients.

However, many organisations do offer guidelines on specific nutrients. The WHO, for example, suggests that carbohydrate should be somewhere between 55 and 75% of a person's daily energy intake, protein 10–15% and fat no more than 10%. The problem for the individual who wishes to adhere to these guidelines is that most foods consist of a mixture of nutrients.

◆ Although not everyone eats the same food, different communities each tend to have a typical diet. For the three types of community shown in Figure 2.2, which one most closely matches WHO guidelines? What comment would you make about the Western diet?

◆ The plantain-eating community's diet most closely matches WHO guidelines, although the intake of fat and protein is a bit low. On the other hand, Figure 2.2 suggests that in Western communities there is too great an intake of fat and protein.

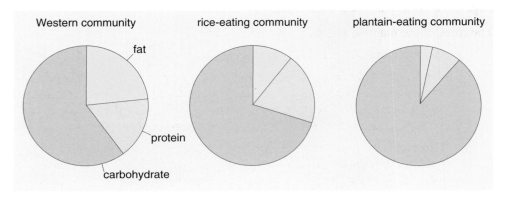

Figure 2.2 Percentage intake of the macronutrients in three types of community characterised by different diets. Plantain is a tropical tree with a banana-like fruit.

Another food group that is not nutritious – i.e. it is not a nutrient – but is important to include in our diet is *fibre*. Fibre (roughage) is non-digestible carbohydrate, but it has an important role in aiding the movement of food through the gut, as you will discover in Chapter 3.

◆ Alcohol is not a carbohydrate, protein or fat, but it does provide energy. Why is it not classified as a nutrient?

◆ Alcohol is not essential for body function (whatever some people may think).

Other components of the human diet are not nutrients at all, as they do not provide energy, nor do they perform any other essential function within the body, but are eaten for other purposes. For example, spices and other flavourings help make food more palatable, and tea and coffee drinks, while providing a good source of water, may also contain other valuable substances such as antioxidants.

2.4 The chemical nature of macronutrients

To describe the structure of the macronutrients, you require knowledge of some of the language used by chemists. Everything around us, and in us, is made up of *atoms*. Atoms are the basic building blocks for everything, from naturally occurring substances such as rocks, trees, copper and gemstones to manufactured articles such as plastics and paper. Some atoms exist individually (helium exists as individual atoms in the atmosphere) but mostly atoms are only stable when joined to other atoms (oxygen exists in the atmosphere as two oxygen atoms joined together). When atoms join together, they are held by forces that are known as *chemical bonds*. Two or more atoms held together by chemical bonds are known as a *molecule*. When different types of atoms join together, chemists describe the substance as a *compound*: water is a compound, made from oxygen and hydrogen atoms. Macronutrients are also compounds that contain oxygen and hydrogen atoms, but they contain other atoms too, such as carbon. They are mostly large molecules so are described as *macromolecules*.

The roles of the macronutrients are outlined below.

- Carbohydrates are usually the main energy source for the body.
- Fats (scientifically known as lipids) are a rich source of energy, as well as being key components of cell membranes and signalling molecules such as those used for sending information between different areas of the brain and other parts of the nervous system.
- Proteins are involved in the growth, repair and general maintenance of the body. Many hormones, which are the signalling molecules carried in the bloodstream, are proteins. Dietary proteins consumed in excess of the need for maintenance can be used to provide energy.

2.4.1 Carbohydrates

Carbohydrates are a large group of compounds and include sugars, starch and fibre.

The simplest form in which a carbohydrate can exist is as a monosaccharide. There are three monosaccharides that are important in the human diet: glucose, galactose and fructose. When two monosaccharides link together, a disaccharide is formed, and when many monosaccharides link together, a polysaccharide is formed (see Figure 2.3a). For example, sucrose (table sugar) is a disaccharide made up of one glucose and one fructose molecule, and lactose (which is found in milk) is made up of one glucose and one galactose molecule.

Derived from the Greek, mono- means one, di- means two and poly- means many; saccharide comes from the Latin for sugar.

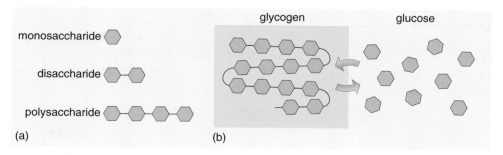

(a) (b)

Figure 2.3 (a) Schematic diagram of a monosaccharide, a disaccharide and a polysaccharide. (b) Schematic representation of the interconversion of glucose and glycogen in the body.

Polymers are chemical compounds formed from repeated units.

Polysaccharides can contain thousands of monosaccharide units. The main polysaccharide found in the human diet is a polymer of glucose called **starch**, which is derived from plants and is found in potatoes and cereals. Animals also have an energy storage polysaccharide called **glycogen**, which is also made up of glucose molecules, although in a different arrangement of units. Glycogen is stored in the liver, kidney and muscle and can be broken down into its constituent glucose molecules when the body requires energy, as illustrated in Figure 2.3b.

Glucose cannot be stored in body cells (other than as glycogen in liver, kidney and muscle and, by conversion to fat, in adipose cells) and the brain cannot use any other energy source in the short term, so the body must ensure that there is a steady supply of glucose in the bloodstream from which brain cells can extract glucose according to their needs. The way in which glucose supplies are maintained and used to provide a continuous source of energy for the brain will be considered in Chapter 4.

2.4.2 Lipids

Lipids are defined as being insoluble in water and encompass fats (solid) and oils (liquid). Unlike carbohydrates, lipids do not form polymers, but aggregates of individual and different lipid molecules can associate together. The lipids that store energy are the **triacylglycerols** (often shortened to **TAGs**), which are made up of three **fatty acid** molecules combined with one molecule of **glycerol**, as illustrated in Figure 2.4a. The fatty acid chains tend to be of different lengths; the longer the chain, the more energy it has stored in it.

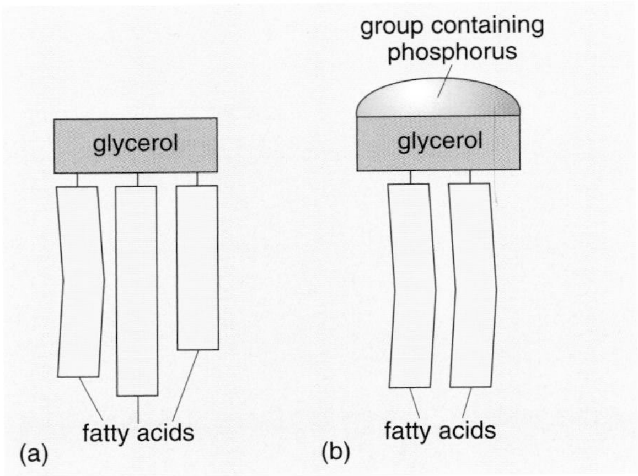

Derived from the Greek, tri- means three.

Figure 2.4 (a) Schematic representation of a TAG molecule. Note how three different fatty acids (shown by different lengths) are attached to this glycerol molecule. One fatty acid is bent in one place; it is described as a monounsaturated fatty acid. (b) Schematic representation of a phospholipid molecule. Note how these are formed from two fatty acids and glycerol together with another group of atoms that includes a phosphorus (P) atom.

◆ What is the name of the tissue that stores the TAGs?

◆ TAGs are found in the adipose tissues of the body.

In everyday speech, adipose tissue is called 'fat', and the definition of obesity uses that term too. You will also find the TAGs being referred to as 'fat'. In the context of obesity, and within this book, it should be clear when the term 'fat' is being used to mean adipose tissue and when it is being used to refer to TAGs.

There are other types of lipids in our bodies, but they are not used as energy stores. Examples include another class of fatty molecules called phospholipids (Figure 2.4b). Phospholipids form the major component of the cell membranes. Cholesterol, a lipid with several vital roles in the human body, is also a component of cell membranes. Additionally, cholesterol is a **precursor** (building block) molecule for the manufacture of other molecules such as the male and female sex hormones (testosterone and oestrogen), vitamin D and bile salts (which are important for digestion, and about which there will be more in Chapter 3).

2.4.3 Proteins

Proteins are a large and diverse group of biological molecules. Each protein is built up of small molecules called **amino acids**, i.e. proteins are polymers of amino acids. Very short chains of up to about 30 amino acids are called **peptides**. Proteins are made within cells using combinations of 20 different amino acids. There are eight amino acids that humans cannot make, so must be taken in as part of the diet. (These are therefore known as essential amino acids.) The many thousands of different proteins each have a particular biological function and structure. Proteins are constantly being synthesised (made) within cells under 'instructions' from genes (Figure 2.5).

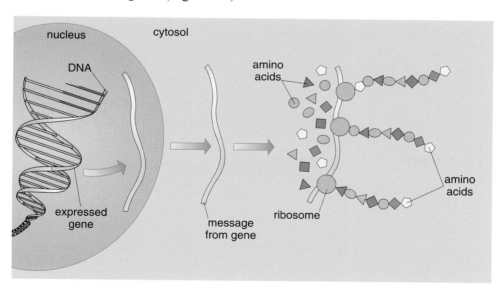

Figure 2.5   Genes make proteins. To allow the gene to be expressed, the long DNA molecule cannot remain packed into the chromosome structure. It unravels enough for the gene to be exposed and be able to make a messenger molecule. This message goes into the cytosol, and organelles called ribosomes attach to it and are able to make the protein specified by the gene.

Different cells will be making different proteins, each of which will have a particular biological function and structure. For example, skin cells make the components to make more skin cells, whereas a salivary gland cell might be making saliva. From your own experience, you know that salivary gland cells vary in the amount of saliva that they produce over 24 hours. It is characteristic of genes that their activity varies over time. When a cell is actively making a protein under instructions from a gene, the gene is said to be being **expressed**. When a gene is not making protein, it is said to be *switched off*.

So it is genes that provide the 'instructions' for amino acids to be linked together to form protein chains. As each chain consists of hundreds of the 20 different types of amino acids, there are thousands of possible sequences. Each particular type of protein molecule has its own unique sequence of amino acids along its length, just as a word has a unique sequence of letters. As in forming words, not all the letters of the alphabet are used, so in any one protein some amino acids may not be used at all whereas others may occur many times. The chain of amino acids is then folded into unique three-dimensional structures appropriate for that protein's function.

Sometimes an error will be found in the genetic material of a new cell and as a result a gene might be unable to specify the correct sequence of amino acids to form its protein. The faulty gene is then said to have a **mutation**. There will be more on the contribution of genetic mutations to the development of obesity in Chapter 6.

Typically, somewhere between 12 and 18% of our total body weight is protein. Most of this is muscle and skin. Other body proteins include hair, nails and hormones. There is a constant process of protein synthesis and breakdown going on throughout the body. This is termed *protein turnover*.

◆ Why do you suppose we need a constant supply of protein? Couldn't the body simply keep recycling the amino acids in its own proteins?

◆ Protein is constantly lost from the body, e.g. via the shedding of skin cells and through bodily secretions (such as saliva, digestive juices and sweat). Loss is also heavy at times of illness and injury, with loss of blood and damaged tissues.

The glands that secrete sweat, saliva and digestive juices are called exocrine glands.

Under normal circumstances, body protein is not used to release energy. However, the body's own muscle tissue will be broken down to provide energy under conditions of starvation. In fact, this process begins before fat stores have been used up. Hence muscle wasting can occur when slimmers are on very restrictive diets (see Chapter 8).

2.5   Measuring energy intake

Your energy intake depends on what foods you eat, but how can you know how much energy each morsel provides? Some processed foods give this information on the packet, but how has this been measured? What about raw food, such as meat and vegetables?

Energy is measured in **kilocalories (kcal)** or kilojoules (kJ): one kcal is approximately 4.2 kJ. Although the kilojoule is the internationally agreed scientific unit for measuring energy, we will use the kilocalorie in this book because that is the term most commonly used by the food industry. (It is, of course, equivalent to 1000 calories, but in discussions of dieting the kilocalorie is shortened to calorie in some parts of the world – so 'calorie counting' is actually 'kilocalorie counting'.) A kilocalorie is defined as the amount of energy required to raise the temperature of 1 litre of water by 1 °C. Scientists and food technicians burn known quantities of a food in oxygen and measure the heat output to gain an indication of the energy content of foods. There are various tables published that give the calorific values of most foods, but they are somewhat approximate. For example, the calorific value of a cake, a biscuit or a stew will depend on the precise mix of ingredients that have been used. Nevertheless, most processed foods have an indication of their calorific value. The heat energy that is released by burning the food in oxygen is equivalent to the energy released in the body when the complex molecules in the food are broken down.

The approximate energy yields of the macronutrients are shown in Table 2.1.

◆ How many times greater is the energy yield of fat than that of protein (in kcal g^{-1})?

◆ Fat yields 9 kcal g^{-1} of energy and protein yields 4 kcal g^{-1} of energy. So fat yields $\frac{9}{4} = 2.25$ times more energy per gram than protein.

g^{-1} is the conventional scientific way of writing 'per gram'; m^{-1} means 'per metre', and so on.

In other words, fats are a more concentrated form of energy. They are more energy-dense than protein and carbohydrate, so it is therefore rather too easy to take in more energy than one requires when eating a high-fat diet, and to discover subsequently that the excess is being stored as fat in the adipose tissues of the body.

In Chapter 8, you will look more closely at the way energy intake can be calculated and related to weight gain. In the next section, you will gain an idea of how energy stored in food is released and is used to drive the metabolic activities of cells.

Table 2.1 Approximate energy yields of the macronutrients.

Macronutrient	Available energy/kcal g^{-1}
carbohydrate	4
fat	9
protein	4

2.6 Releasing energy from food

Energy is needed in every body cell. *Mitochondria*, the oval-shaped structures in Figure 1.5a, are often called the power houses of the cell because it is within them that a molecule called **adenosine triphosphate (ATP)** captures energy released from nutrients such as sugars (e.g. glucose) and fats (e.g. fatty acids) by the process of **cellular respiration**. Cells that are very active, such as muscle cells, will contain many mitochondria. For glucose, cellular respiration can be summarised in the following word equation as:

glucose + oxygen $\longrightarrow$ carbon dioxide + water + energy

ATP can be thought of as an 'energy currency' (like money) which can be made and then spent to get things done. Unlike money, it cannot be saved. Each ATP molecule typically exists for only a very brief period – a matter of seconds – before being utilised to provide energy for processes such as the synthesis of new macromolecules – for example, proteins.

We can write ATP where previously we had written energy:

glucose + 6 oxygen + 36 ADP + 36 P_i $\longrightarrow$ 6 carbon dioxide + 6 water + 36 ATP

This tells us that a single molecule of glucose (on the left side of the equation) will produce 36 molecules of ATP (on the right), provided that 6 molecules of oxygen, 36 molecules of ADP (adenosine diphosphate) and 36 molecules of inorganic phosphate (P_i) – one of the minerals your diet needs to contain – are available with the glucose.

Although the equation above shows glucose as the provider of energy, all three macronutrients (carbohydrates, fats and proteins) can be used to provide energy, as can be seen if you look at the upper portion of Figure 2.6. This figure shows *biochemical pathways* (*metabolic pathways*) in the cell. It indicates that energy is not released explosively in one fireball, as when sugar is set alight. Instead, energy is released gradually through a series of chemical transformations that involve the breaking and re-forming of chemical bonds, and the rates at which these occur are carefully controlled by **enzymes**. Enzymes are large protein molecules that have the ability to accelerate a *particular* chemical reaction in a cell. Enzymes remain unchanged at the end of the reaction. There are in fact many more tiny steps in these metabolic pathways than are shown in Figure 2.6.

◆ Why do you suppose that energy-containing food molecules are gradually broken down through a series of small steps that yield molecules of ATP at several stages?

◆ A sudden massive energy release, as occurs with burning, would destroy the cells of the body.

Figure 2.6 makes the important point that whether glucose, fatty acids, glycerol or amino acids are used to provide energy, they all enter into common biochemical pathways in the cell. However, they do not all yield the same amount of energy (a fact that cannot be deduced from Figure 2.6). Under ordinary circumstances, carbohydrates form the major component of most people's diets, and therefore most energy is obtained from carbohydrates.

◆ Although most of your energy is likely to be derived from carbohydrates, which nutrient type provides the most energy per gram?

◆ Fat provides the most energy per gram (see Table 2.1).

(You might also have reflected that although Table 2.1 gives a single value, it must be an 'average' value because TAGs with long fatty acid chains release more energy than those with shorter fatty acid chains (Section 2.4.2).)

Box 2.1 Metabolic pathways to ATP production

Food is broken down in the gut (shown at the top of Figure 2.6), a process which will be described in Chapter 3. The molecules that can be used by cells to produce ATP are glucose, amino acids, fatty acids and glycerol. These molecules leave the gut and are circulated around the body in the bloodstream. All body cells can withdraw from this resource the fuel molecules and the oxygen they need to produce ATP. Most of the cell's ATP is made in the mitochondria, but some ATP is made in the watery cytosol as a result of the metabolic reactions that convert molecules of glucose and glycerol into a chemical compound called pyruvate. Amino acids can be converted into pyruvate too, but if you follow the metabolic pathway between amino acids and pyruvate you will note that no ATP is formed. Pyruvate is transported into the mitochondria, where it is converted into acetyl CoA. Fatty acids and amino acids can also be converted into acetyl

CoA. In fact, fatty acids are converted to acetyl CoA in a number of chemical transformations, requiring oxygen and glucose, that generate a great deal of ATP. Acetyl CoA enters a pathway called the TCA cycle, where more ATP is made. Waste carbon dioxide is also produced at this stage (as is produced from the earlier conversion of pyruvate into acetyl CoA). Other products of the TCA cycle are further processed to make more ATP, and this step requires oxygen and produces water as a waste product.

This is not a diagram to memorise! The points of interest are that carbohydrates, fats and amino acids, despite being very different, can all be metabolised to produce ATP and that they all finally enter common pathways to produce ATP. Also, these energy-producing pathways are busy in all your cells all the time – although activity is not equally frenetic in all areas at all times.

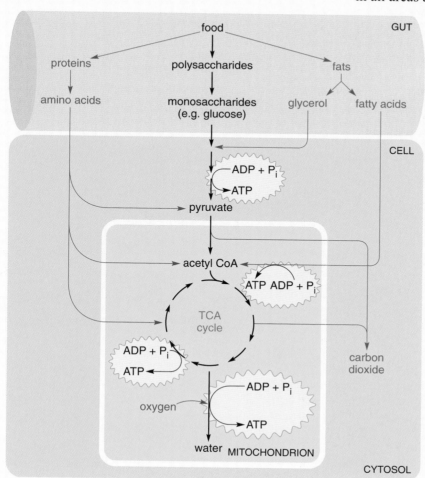

Figure 2.6 Summary of the fates of proteins, carbohydrates and fats in metabolic pathways that result in the production of ATP. Mitochondrion is the singular of mitochondria.

Although the energy released when nutrients are metabolised in the body is immediately used as the ATP molecules provide energy to enable other chemical reactions and cellular processes to proceed (i.e. it enters other biochemical pathways), a lot of energy (60–75%) is lost as heat. This heat maintains the body temperature. The liver produces more heat than any other organ because, as you will discover in the next two chapters, it is continuously, metabolically very active.

◆ Which other cells would you expect to use large amounts of energy and generate a lot of heat?

◆ Muscle cells have high energy requirements when individuals are physically active and you will know how hot you can become if you are being very energetic! You might have added that muscle cells 'shiver' (vibrate) to produce heat when you are cold – for example, after a swim in cold water. Although muscle cells can produce a lot of heat, their activity is intermittent.

2.7 Energy expenditure

When thinking of energy expenditure, you might have imagined that physical activities such as walking and active sports would be the main users of energy. Although physical activity is obviously energy-dependent, it is not the main user of energy in the majority of people. The major component of TEE is the **basal metabolic rate (BMR)** (sometimes called the resting metabolic rate), which is the amount of energy required to carry out the basic processes of life – the processes that continue as you sit quietly or lie asleep, such as breathing and the beating of your heart. BMR values are remarkably constant when related to lean body mass – body weight (mass) minus the weight (mass) of the fat tissue. The greater the lean body mass, the greater the BMR. People who are obese have a higher BMR because they have an overall greater body mass, i.e. both their lean and fat body mass are greater than those who are of healthy weight. Of course this means that as a person loses weight, their energy needs decrease; this lowering of BMR is just one of the many small ways in which the body's metabolism opposes efforts to become slimmer! BMR also changes with the age of the individual and differs between the sexes. It declines with age due to a decline in lean tissue mass. So older people have a decreased energy need and should reduce energy intake (eat less) if they want to avoid putting on weight.

◆ If comparing a man and woman of the same height and weight, who would you expect to have the higher BMR and why?

◆ You would expect the man to have the higher BMR because he has the greater proportion of lean tissue (see Section 1.3.3).

It is estimated that in most people BMR consumes 60–70% of TEE if they are leading fairly sedentary lives. BMR increases in pregnancy and lactation (see Chapter 6) and also when the body is diseased. (Metabolic rate increases by about 8% for every 0.5 °C increase in temperature, e.g. during a fever.)

When we eat, our energy expenditure increases and is known as diet-induced thermogenesis (DIT). The processes involved in digesting food, about which there will be more said in Chapter 3, take a lot of energy. The admonition not to eat just before taking exercise is based on the premise that the body will be stressed by the competing demands for energy. (Also, the food entering the stomach can stretch the stomach and put pressure on other organs in the vicinity.) Over 24 hours, the work required to digest meals can account for as much as 10% of TEE.

◆ If you were to reduce the amount of food you ate over 24 hours, what would happen to the amount of energy expended?

◆ If you eat less food, you reduce your absolute energy expenditure (i.e. you reduce DIT).

This has repercussions for anyone trying to lose weight by eating less. When you eat less, the amount of energy you need for DIT decreases, thereby diminishing the effect of your efforts. This is a fairly small effect, but it is unhelpful.

Further energy expenditure is necessary to carry out everyday activities beyond the energy required for the BMR and DIT. Physical activity (such as walking) is the second highest user of energy at around 20–40% of TEE in people who are relatively sedentary. The value is noticeably different in very active individuals, as can be seen in Vignette 2.1 (overleaf). Obese people have higher energy expenditure when they undertake physical activity because they have a greater body mass to move. One small comfort for them: this time the body's conformation is working in their favour! However, as they successfully lose weight, they will lower their energy expenditure if their level of physical activity is unaltered, so can compensate for this by increasing the time and/or intensity of their exercise. The evidence is that there are many obstacles to taking exercise for overweight and obese people (see also Chapters 7 and 8), but that as they lose weight and improve their physical fitness it becomes easier and more pleasurable to undertake physical activities.

The amount of physical activity, apart from formal exercise, can vary considerably even between individuals who apparently lead very similar lives. Running rather than walking up stairs and fidgeting rather than sitting still can make big differences to the amount of energy used.

Table 2.2 shows estimates of energy expended during different activities for a 70 kg non-obese person.

Table 2.2 Estimates of energy expenditure per hour during different activities for a 70 kg non-obese person.

Form of activity	Energy expended/kcal h^{-1}
lying still, awake	77
sitting at rest	100
typing rapidly	140
dressing or undressing	150
walking on level at 4.8 km h^{-1}	200
jogging at 9 km h^{-1}	570

Vignette 2.1   Richard, Ben and their levels of physical activity

Figure 2.7   Richard out training.

Richard (Figure 2.7) is in his twenties and, in addition to training for a career as a physiotherapist, he cycles competitively. Competitive cycling makes huge energetic demands and he has to plan ahead to ensure that these demands are met. For example, when he is in training and cycling 3 hours a day, he needs to ensure an intake of 5000 kcal; when he cycles 6 hours a day, this has to go up to 7000 kcal. Richard points out that cycling is one of the few sports where eating is compatible with exercising – in fact, on the long rides it is essential to eat or he has insufficient energy to make the final burst to the finishing line. The kinds of foods he uses are liquid energy drinks and concentrated-energy cereal bars. Richard says he finds it hard to eat enough when he is in training and he also uses a body monitor to make sure that his body fat does not drop below 11%. His BMI is usually a healthy 22–23.

Richard and his friend Ben notice a story in the local newspaper about a young man of a similar age to themselves who has a genetic disorder called Prader–Willi syndrome that causes compulsive eating. The paper reports that this young man consumes 5000 kcal a day, at a cost of £9000 a year, and now weighs 40 stone and is unable to move out of his chair. Richard remarks that this dramatically demonstrates how lack of exercise causes obesity. Ben is not so sure. He points out that, unlike Richard, he has never been keen on exercise. He left school to work as a scaffolder 6 years ago and the two young men have remained exactly the same size, even able to borrow each others' clothes when going out.

A year after this conversation, Ben got a job driving buses and put on 2 or 3 stone (13–19 kg) rapidly, despite not changing his eating habits or his (almost non-existent) exercise habits. Richard tells Ben that 'exercise' does not have to be a formal sporting activity and Ben is now convinced that as a scaffolder he was actually taking a great deal of exercise. Ben started to exercise at the local gym and returned to a scaffolding job. Within a year he had lost all the weight he had gained as a bus driver.

2.8   Controlling weight

You are now in a position to consider the problems that face every individual in relation to keeping to a comfortable and static weight. The only way this can be achieved is by exactly balancing energy intake and energy expenditure. Yet most people live varied lives from day to day; varied in the types and amounts of food they eat and varied in the amount of activity they undertake. In this section, we consider whether it is possible that, just as activities such as breathing and temperature regulation occur without conscious thought, the body has mechanisms that enable it to maintain a stable weight without the individual having to think about what they are doing.

2.8.1   Homeostasis

As mentioned in Section 1.7, the body has a remarkable ability to regulate its activities without any conscious input being made. Important variables critical for life are held within limits and are said to be *regulated variables*. A well-known example is body temperature. An often-used measure for determining whether someone is unwell is to 'take their temperature' using a clinical thermometer. The thermometer is marked at 36.5 and 37.5 °C, and if the temperature reading falls between these two marks, the person is deemed to be well as their body temperature is being maintained at around 37 °C. This level allows normal physiological activity, i.e. the body systems involved in maintaining temperature are functioning normally. (**Physiology** is the study of the relationship between structure and function of body systems.) The normal functioning involves a self-regulating mechanism whereby any deviations from optimum conditions in the body tend to cause responses that return the system to the optimum.

Staying with the temperature example, when our body temperature falls, we start to shiver. This comes about because there are *neural* mechanisms mediated by specialised neurons (nerve cells) that detect the drop in body temperature and signal this information to the brain. The brain signals to muscles to start shivering, and the heat generated by these muscle movements will tend to return body temperature to normal. In the 1930s, the US physiologist Walter Cannon coined the term **homeostasis** to describe the various physiological systems which serve to restore the normal state, once it has been disturbed.

Homeostasis is an essential biological principle with two important and related aspects:

- life is only possible provided that certain key variables of the body are maintained within limits
- deviations from these optimum conditions in the body tend to cause responses that return the system to the optimum.

The type of response that we examined above (shivering in response to a fall in temperature) is an example of a **negative feedback** response. Negative feedback is the process by which a control mechanism reacts to a change in the output of the system by initiating a restoring action (Figure 2.8). Negative feedback systems maintain a preset state, so they are stable and an important feature of homeostasis.

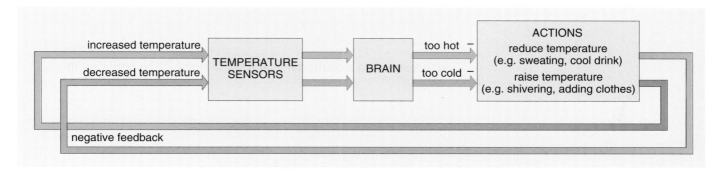

Figure 2.8   The principle of negative feedback, with body temperature as an example. The brain's action is based on a comparison of the actual body temperature (from information provided by the temperature sensors) against an optimum value, which is set by the brain. If body temperature is not optimum, the brain then triggers action that will result in a body that is too hot being cooled (e.g. by sweating) or a body that is too cold being warmed (e.g. by shivering). These subsequent changes are *fed back* to the brain, *negating* the effects of the original, non-optimum situation and restoring body temperature to the optimum value. (The minus sign (−) represents this *negative* feedback in action, i.e. ongoing actions are inhibited.)

Although negative feedback is crucial in re-establishing normal conditions when deviations occur, it is not the only homeostatic mechanism involved in the maintenance of the body's internal environment.

◆ Think about your own experiences and try to recall some examples of where homeostasis is maintained by your *behaviour*, even though no deviations from normal have yet happened.

◆ In winter you might put on warm clothes before leaving your centrally heated home. You may eat breakfast even though you do not feel particularly hungry.

Such anticipatory actions are, in effect, a process that is termed **feedforward**, to distinguish it from feedback control where the response is to a disturbance that has already occurred. The above examples of feedforward are forms of behaviour that we perform in full consciousness of their effects and with this purpose in mind. In fact, there are other, involuntary feedforward mechanisms which also play an important role in homeostasis. You will learn more about these, particularly in Chapters 4 and 5.

Many of the regulated variables, such as body temperature, that are defended by homeostatic mechanisms are held nearly constant, regardless of circumstances. So it may seem strange to be suggesting that body weight is also a regulated variable, given the very substantial variation that we can observe both over the course of a lifetime and from one person to another. However, it is often argued that the relatively small degree of individual variation in body weight over short time periods implies that there must be some regulation. In addition, as you will see later in this book, there is a series of hormonal and neural mechanisms influencing energy intake that have many of the characteristics of a homeostatic system. Of course, body weight may also change – perhaps permanently – with changes in the environment. A plentiful supply of highly palatable food will increase the likelihood that humans (and their household pets!) become obese. However, even in this case, the typical pattern is for weight to increase and then be held, i.e. plateau at a new and higher level. Thus for body temperature it seems realistic to talk of a set point, even if we cannot identify its location in the body, whereas for feeding we often talk in terms of a sliding set point or a settling point. Despite this cautionary note, it still seems appropriate to discuss obesity in terms of failure of – or disturbance in – a homeostatic system.

2.8.2 Why homeostasis fails

How might automatic body weight regulation be achieved? To answer this question, you need to know more about eating behaviour (Chapter 5) and how food is digested and nutrients absorbed (Chapter 3) and then metabolised to provide energy or moved into storage (Chapter 4).

If you make the assumption that there is a homeostatic mechanism at work, but that it is currently failing to keep on top of the situation in many societies, then the following activity should help you to identify some possible reasons why this homeostatic mechanism is currently failing.

Activity 2.1   What does my body tell me?

Allow 10 minutes for this activity

In relation to energy intake, list any regulating body signals to which you respond: what makes you start eating and what makes you stop?

In relation to energy expenditure are you aware of any regulating body signals to which you respond?

Now look at the comments on this activity at the end of this book.

From the lists you generated in Activity 2.1, were there any signals that might predispose you towards achieving a positive energy balance? In other words, why might someone living in a modern urban environment eat more than they need and/or fail to utilise their energy intake to avoid gaining weight?

If you look at our list at the end of this book, the underlying factors that appear relevant are:

- ready availability of highly palatable food

- the fact that for many people to get from A to B does not require much physical energy because they use some other form of transport (Figure 2.9).

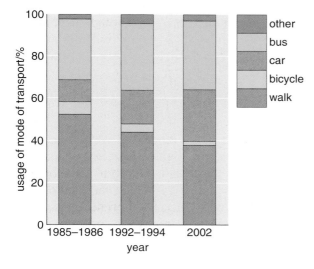

Figure 2.9   Modes of transport used by children aged 11–16 to get to school in the UK: trends over three decades, 1985/6–2002.

In Chapter 8 you will look in more detail at the effect upon a person's energy balance that could result from making the change from taking a bus to walking to school. You will also examine the effect of making a small but consistent reduction in eating habits, such as removing a packet of crisps from a lunch box. Meanwhile, study the graph shown in Figure 2.10, which shows the increase in the number of 'fast food' outlets opening in Singapore over the period from 1975–1995 and the prevalence of diabetes during that same period.

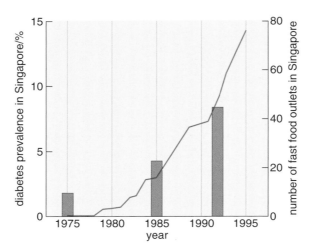

Figure 2.10   The number of 'fast food' outlets in Singapore over the period 1975–1995 (line) and the prevalence of diabetes (bars).

◆ Describe the relationship between the two variables in Figure 2.10.

◆ The Singaporean population was not free from diabetes in 1975 before the first 'fast food' outlet opened, but the increase in the number of people with diabetes follows the increase in the number of 'fast food' outlets.

Where one variable increases (in this case, the number of people with diabetes) as another variable increases (in this case, the number of 'fast food' outlets), there is said to be a **positive correlation** between the two variables.

You might be tempted to speculate that the increased availability of 'fast food' leads to people eating a less healthy diet and putting on weight. It is certainly true that obesity is an environmental risk factor for diabetes, but Figure 2.10 does not provide *evidence* for a causal relationship.

Indeed, it is true that much of the evidence implicating the obesogenic environment in the increase in the numbers of people who are overweight or obese is based on correlations such as that shown in Figure 2.10. Nevertheless, when you have read more about the way in which your body metabolises and responds to food, the evidence for the connection will become overwhelming.

2.9   Summary of Chapter 2

2.1   If energy intake exceeds energy expenditure, the body has a positive energy balance and will gain weight.

2.2   Energy is provided by the macronutrients – fat, carbohydrate and protein – in food. The WHO guidelines suggest that carbohydrate should provide 55–75% of a person's daily energy intake.

2.3   Glucose (from carbohydrate), fatty acids and glycerol (from fat) and amino acids (from proteins) are the nutrient molecules that can provide energy for the body's metabolic activity.

2.4 Fats yield more energy per gram than either carbohydrates or proteins.

2.5 Energy is released from nutrients within individual cells when and where it is needed. The energy is captured by a molecule called ATP, which almost immediately transfers this energy into other biochemical pathways.

2.6 For people leading relatively sedentary lives, the major part of their energy intake will be spent on general maintenance (staying alive), scientifically known as their BMR (basic metabolic rate). The absolute level of BMR is higher in obese individuals because they have a greater body mass to maintain.

2.7 Energy associated with digestion (DIT) can consume up to 10% of total energy expenditure (TEE). DIT expenditure will decrease if less food is consumed, e.g. when trying to lose weight.

2.8 Except in very sporty people, physical activity uses less than 40% of TEE.

2.9 Homeostasis describes the body's ability to hold many variables at a steady value, despite disturbances that alter these values.

2.10 Body weight does remain remarkably constant – in the short term, at least – and it is suggested that the development of obesity represents a failure of some of the body's homeostatic systems.

2.11 It might be that the body's homeostatic systems are failing under increasing pressure from an obesogenic environment.

Learning outcomes for Chapter 2

LO 2.1 Define and use, or recognise definitions and applications of, each of the terms printed in **bold** in the text.

LO 2.2 Explain the link between energy balance and body weight.

LO 2.3 Outline the role of macronutrients, particularly in relation to providing energy.

LO 2.4 Outline the role of ATP in cell metabolism.

LO 2.5 Describe how energy expenditure relates to a variety of factors, including body weight.

LO 2.6 Explain why the development of obesity is regarded as a failure of homeostatic mechanisms.

Self-assessment questions for Chapter 2

Question 2.1 (LOs 2.1 and 2.2)

Re-read Vignette 2.1 and state when Ben moves to a state of negative energy balance.

Question 2.2 (LO 2.3)

Is it true to say that amino acids are stored in the body as proteins, so that they can provide energy if neither carbohydrates nor fats are available? Explain your answer.

Question 2.3 (LOs 2.4, 2.5 and 2.6)

Which of the following statements about metabolism are true?

(a)   Energy can be stored for about 24 hours in any body cell as ATP.

(b)   Proteins are made from ATP.

(c)   The energy from food in the diet cannot be extracted without an input of energy.

(d)   The fact that body weight does not fluctuate dramatically in the short term suggests that body weight is a regulated variable.

CHAPTER 3

FOOD: DIGESTION AND ABSORPTION

Dis-moi ce que tu manges, je te dirai ce que tu es.

Anthelme Brillat-Savarin, French lawyer, politican and gourmet, 1755–1826

3.1   Introduction

The idea that there is a strong link between what you eat and your health and wellbeing has a long history. As Brillat-Savarin observed in his book, *Physiologie du goût* (*The Physiology of Taste*), 'Tell me what you eat and I will tell you what you are'. It is supposed that this is the origin of the saying 'you are what you eat'.

Have you ever, on unexpectedly meeting a friend while shopping in the supermarket, wished you could hide some of your intended purchases? Perhaps you might feel that the text could be rewritten as 'look into my shopping trolley and you'll see what I'm going to become'!

In the previous chapter, we identified the components of food that provide energy that, on the one hand, powers all body activity and, on the other hand, is converted into stored fat if consumed in excess of TEE.

The energy-providing foods – fats, carbohydrates, proteins and alcohol – are a mix of molecules, the majority of which are macromolecules. These macromolecules are insoluble in water and too large to be taken up by cells, so must be broken down into smaller, soluble subunits, which can be taken up from the gut into the bloodstream and then delivered to the different parts of the body. These processes involve the action and interactions of several different organs (e.g. salivary glands, stomach, liver) which are collectively known as the *digestive* or *gastrointestinal system*.

An organ is formed from different tissues that collectively perform a common function.

A number of distinct physiological activities are involved in the assimilation of the nutritious components of foods into the fabric of your body. The processes by which the components of food are broken down in the gut into simpler forms are collectively known as **digestion**, and these can be chemical or mechanical. The uptake into the body of the products of digestion and of small molecules that do not need to be broken down (such as water, alcohol, mineral salts and vitamins) is known as **absorption**. Special **secretions** – substances made in (and then exported from) cells – are needed to digest foods; important constituents of these secretions include enzymes that break the chemical bonds linking the components (i.e. subunits) of macromolecules together. The mixing of ingested foods with digestive secretions in the stomach and the propulsion of intestinal contents along the length of the gut are achieved by intestinal movements, which are referred to as gastrointestinal motility. The elimination of faeces (waste matter together with unabsorbed materials) is known as defecation.

Gastro- is from the Greek word meaning related to the stomach.

In this chapter, you will learn about the digestive system and about the physiological processes that are concerned with the digestion and the absorption of food. This will enable you to understand better some of the methods, discussed in subsequent chapters, that are used to try to prevent the development of obesity.

3.2   The digestive system

The gastrointestinal or digestive tract, often simply termed the *gut*, is essentially a very long tube, the shape and dimensions of which vary according to the physiological activities that take place in each particular region. The different parts of the gut are given different names, some of which will probably be familiar to you. The anatomy (structure) of the gut is shown in Figure 3.1. Associated with the gut are a number of other organs that, by producing digestive secretions, play an essential role in digestive processes. These organs, such as the liver and pancreas, are also shown in Figure 3.1. Note in Figure 3.1a how the intestines are folded into the area known as the *abdominal cavity*; you can observe that they are a tight fit! This is the area that expands when the visceral fat depot expands, leading to central obesity (Section 1.2). Figure 3.2 is a medical scan of the abdominal area, in which you can see the amount of space being occupied by the white fat and imagine how this would push the guts forward, leading to the development of a paunch.

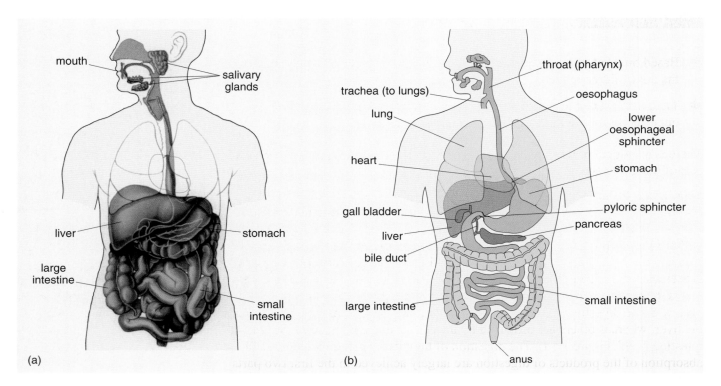

(a)   (b)

Figure 3.1   Anatomy of the digestive tract (gut) and associated organs. (a) View of some internal organs within the abdominal cavity; overlying bone and other tissue has been removed. (The structures shown in outline on either side of the oesophagus are the lungs. They lie in the thoracic cavity, together with the heart – also shown in outline.) A tough sheet of tissue called the diaphragm separates the abdominal cavity and the thoracic cavity. (b) A simplified diagram of the gut and associated organs.

The gut is very long; in an average adult, it measures about 7 m in length. A long gut is characteristic of other primates (e.g. monkeys, chimpanzees and gorillas) who eat a largely vegetarian diet that requires a lot of digesting to extract the available nutrients. In other words, our gut is very efficient at gaining maximum value from the food that is available. If this efficiency were reduced, less energy would be delivered to the body.

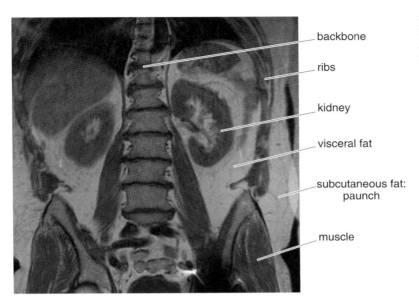

backbone

ribs

kidney

visceral fat

subcutaneous fat: paunch

muscle

Figure 3.2 MRI scan showing abdominal fat. This shows up as white around the darker kidneys, muscle tissue and bone.

◆ Based on what you have just read, state one way in which the efficiency of the gut could be reduced.

◆ The gut would be less efficient if it were shorter, and thus offered less opportunity for the digestion and absorption of nutrients.

There is a surgical procedure that you will read about in Chapter 9 that bypasses part of the gut to achieve this effect.

Food enters the gut through the mouth, which is involved in the initial processing of food. The mechanical chewing and the lubrication of each mouthful of food with saliva aids swallowing and eases the passage of the *bolus* (the soft, chewed ball of food) through the throat and, via the **oesophagus**, to the **stomach**. The stomach is a muscular, bag-like organ where the first major digestion and mixing processes occur. The partially digested food then passes into the **small intestine**; this is the longest part of the gut and is where most digestion and absorption occurs. Some digestive secretions enter the small intestine from the pancreas and the liver, whereas others are produced by the cells lining the inside of the small intestine itself (Figure 1.5d). The digestion of the dietary macromolecules and the absorption of the products of digestion are largely achieved in the first two parts of the small intestine (called the *duodenum* and *jejunum*); the last part (the *ileum*) provides additional capacity. Absorption of water and mineral salts does occur in the small intestine, but mainly occurs in the large intestine.

From the small intestine, food passes into the **large intestine**, so called because it is wider than the small intestine. The first part of the large intestine is the *caecum*, which is a pouch-like structure from which the very narrow appendix protrudes. The caecum merges with the major region of the large intestine, known as the *colon*. In the caecum and colon, further absorption of water and mineral salts occurs, concentrating the remains of unabsorbed food and waste products into a semi-solid form called faeces. The colon is also the home of millions of bacteria that can metabolise soluble fibre – mucilages from pulses and oats and pectins

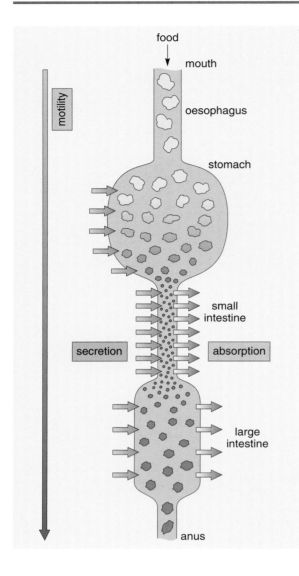

food

mouth

oesophagus

stomach

motility

secretion

small
intestine

absorption

large
intestine

anus

Figure 3.3 The processes occurring in the different parts of the gut.

from fruit – into fatty acids, and gases such as hydrogen and methane. The fatty acids can be absorbed and then used by colonic cells as an energy source, and the gases are lost as flatulence. Some of these bacteria can synthesise vitamins (for example, vitamin K), which are absorbed into the body. Non-soluble fibre, such as the fibres and cell walls of fruit and vegetables, cannot be metabolised. But this roughage has a number of benefits, including delaying stomach emptying. Having a full stomach promotes a sensation of being well fed (**sated**) and can stop you from eating more food. Once the fibre leaves the stomach, its bulk encourages gut motility, and keeping the food on the move prevents the unpleasant sensation of constipation. It may also reduce the risk of certain types of gut cancers – for example, by binding toxic molecules to the fibre and thereby encouraging their removal in the faeces before they can damage gut cells.

The last region of the large intestine is the *rectum*, in which faeces remain until they are expelled by passage through the *anus*. These processes are summarised diagrammatically in Figure 3.3.

3.3 The fate of food as it passes along the gut

Now you have a general picture of the organisation and functions of the gut, we can turn to how the digestive system actually processes foods. Perhaps the easiest and most logical way to describe this is to follow what happens when we eat a meal.

◆ Most of the foods we consume consist of insoluble macromolecules, large molecules that are too big to leave the gut until they are broken down (digested) into their smallest constituents. Name these small molecules.

◆ Amino acids, monosaccharides, fatty acids and glycerol.

These components are displayed more memorably in Table 3.1. These smaller molecules can, with the aid of specialised molecules and processes (to be described shortly), cross the membranes of the cells that line the gut to gain access to the body's circulatory systems and so be transported around and assimilated into the body.

Table 3.1 The macromolecule components of the human diet and their small molecule components.

Macromolecule	Small molecule components
protein	amino acids
carbohydrate	monosaccharides (sugars)
lipids (not strictly macromolecules, but aggregates)	fatty acids and glycerol

As food passes through the gut, it stimulates the release of gut hormones. Over 20 different gut hormones have been identified, but in some cases their precise modes of action and relative importance are yet to be fully investigated.

◆ How can a hormone have more than one mode of action?

◆ Hormones bring about changes by activating receptors on target tissues. If they interact with more than one type of receptor or target tissue, they can have more than one mode of action.

Cholecystokinin (CCK) is a hormone that is released into the bloodstream by duodenal cells in response to food entering the duodenum from the stomach. It stimulates the pancreas's production of pancreatic enzymes and it stimulates the gall bladder (Figure 3.1b) to contract, releasing bile juices into the duodenum. It may also inform the brain of the presence of food in the gut and thereby reduce appetite (see Chapter 5). Other hormones released by the small intestine that seem to have *anorexic* (appetite-reducing) effects are:

- **glucagon-like peptide 1 (GLP-1)**
- **peptide YY (PYY)**.

There will be more about these hormones and how they might be of use in preventing obesity in Chapter 5.

3.3.1 The mouth

Before it reaches the main part of the digestive system, material that is to be digested must be bitten, chewed and swallowed. Although human diets are varied, most consist of at least a proportion of bulky material, such as vegetables, and tough material, such as meat. For food to be digested, it must be in contact with the digestive secretions that act on it, so the bulky food material must first be broken down into small pieces; this process serves to increase the surface area of the food, thereby making it more accessible to the enzymes in the digestive secretions.

During chewing, the food is broken into smaller pieces and mixed with saliva, produced by the salivary glands. Saliva consists of a watery solution, its two main components being an enzyme called *salivary amylase*, which begins the digestion of starch, and *mucus*, which acts as a lubricant. (You do not have to remember the names of the different enzymes, but it is important to note that the individual digestive enzymes cannot break down all food types – only a specific compound.) So salivary amylase breaks down starch but it has no effect at all on fats or proteins.

As we start to eat, sensory neurons in the mouth are activated in response to the taste and feel of food. These activate neurons in the brain which, in turn, activate the neurons *innervating* (supplying) the salivary glands, and so stimulate the secretion of saliva (Figure 3.4). Salivation (the secretion of saliva from salivary glands) is also stimulated in response to the sight and smell of food (and even by the thought of food) as a consequence of a learned association having been formed. This type of learning is known as **conditioning** (see Box 3.1).

Figure 3.4 Schematic representation of the sensory nerves from the mouth, the eyes and the nose, carrying sensory messages to the brain, and the activating nerves carrying action commands to the salivary glands and the stomach.

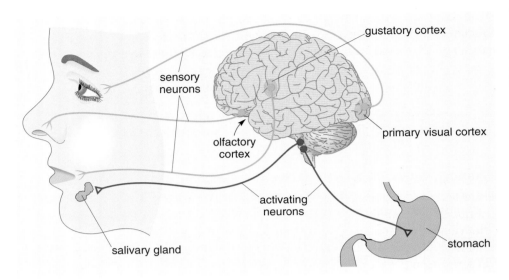

Box 3.1 Conditioning: Pavlov's dogs

The Russian physiologist Ivan Pavlov (1849–1934) was interested in the amounts of saliva that were produced in different circumstances. His dogs salivated at the sight or the smell of meat, although they produced much more saliva once the meat was in their mouths. While conducting a series of experiments on salivation, he noticed that his dogs started to salivate whenever they saw his assistant who brought the meat in to the experimental room – even when he was not carrying any meat. Pavlov deduced that the assistant had acquired salience (meaning) by association with the meat. Pavlov tested this

hypothesis (idea) with a series of experiments where he associated food with a totally unrelated cue, such as the ringing of a bell. So for a period of time, a bell was rung a second or two before the dog was fed. Subsequently, he found that the dog would salivate when the bell was rung, even though no food was being offered. Prior to the pairing with food, the sound had no capacity to trigger salivation. After the pairing with food, the bell acquired the capacity to trigger salivation (Figure 3.5). This capacity was *conditional upon* the pairing with food, and hence arose the term 'conditioning'.

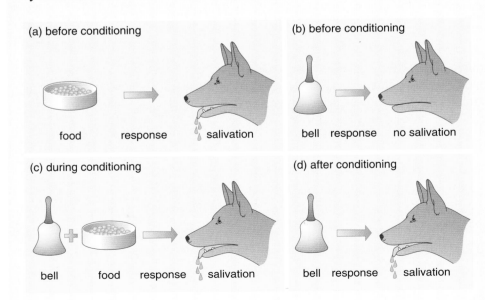

(a) before conditioning

food → response → salivation

(b) before conditioning

bell → response → no salivation

(c) during conditioning

bell + food → response → salivation

(d) after conditioning

bell → response → salivation

Figure 3.5 Pavlov's dog: a schematic representation of the stages in conditioning.

The sense of taste also triggers the secretion of digestive juice in the stomach. The senses of smell and taste often decrease with age, the loss of taste being associated with the loss of taste buds. This helps explain why some older people can start to have symptoms of malnutrition as their taste wanes and they lose interest in eating. In Chapter 5, you will see how the taste or *palatability* of food can affect the amount of food eaten.

3.3.2 The oesophagus

Once swallowed, food passes into the oesophagus and seconds later reaches the stomach, having passed through a sphincter (a ring of thick muscle at the junction of oesophagus and stomach; see Figure 3.1b). Except when food is entering the stomach, this sphincter is closed, but it can open inappropriately in some people, resulting in *reflux* (return) of the stomach contents, known in its mild form as *heartburn*. Too much reflux of stomach contents back into the oesophagus can cause inflammation (swelling and soreness) of the lower end of the oesophagus. This can cause significant discomfort when swallowing and can be relieved by antacids (acid-neutralising compounds). In more severe cases of reflux, drugs which stimulate oesophageal motility can be prescribed, as this should help to push food into the stomach and keep the oesophagus empty. The most important type of gut motility is a rhythmic movement called **peristalsis**: waves of muscle movement propelling food through the gut, as illustrated schematically in Figure 3.6.

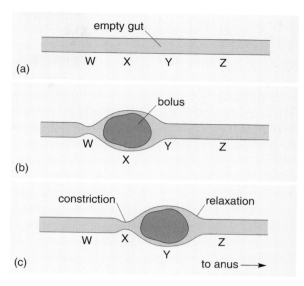

Figure 3.6 Peristalsis. Muscles contract behind the food, but relax ahead of it, propelling the food along the length of the gut: (a) muscle layers relaxed; (b) contraction at W and relaxation at X followed by (c) contraction at X and relaxation at Y results in movement of the food towards the anus.

◆ Obese people often have reflux problems and are advised to use pillows to elevate their head during sleep. In what way will this help?

◆ If there is a slope between the oesophagus and the stomach, gravity will help

3.3.3 The stomach

The stomach undergoes great changes in size, varying in volume from 50 ml (millilitres) when empty to 1500 ml (1.5 litres) after a large meal, i.e. increasing its size 30 times. One method of treating obesity can be to surgically reduce the size of the stomach (see Chapter 9). The processing of food in the stomach is achieved by mixing the contents with gastric juices (digestive secretions produced by cells in the wall of the stomach). The presence of food in the stomach triggers the secretion of gastric juices. However, observation of the timing of such secretions show that, rather than responding to the presence of food in the stomach, the gastric juices start to be secreted before the food has even got into the stomach. In effect, the arrival of food is anticipated.

◆ What is the stimulus that triggers this release?

◆ The taste of food (see above).

◆ What is the name given to anticipatory homeostatic processes such as this?

◆ Anticipatory homeostatic processes are feedforward systems (see Section 2.8.1).

Contractions of the muscular stomach mix the lumps of food with the gastric juices. Gastric juices contain a number of substances, including a strong acid (hydrochloric acid) that inactivates salivary amylase, preventing further breakdown of starch for the time being. Hydrochloric acid dissolves the lumps of food, forming a thick, soup-like mixture called *chyme*. Chyme allows easier access for digestive enzymes, which would otherwise not penetrate lumps of food effectively. The gastric juices contain the enzyme *pepsin*, which digests protein (but none of the other macronutrients), and the enzyme **lipase**, which digests lipids (and only lipids). However, for reasons that will be explained in Section 3.3.4, lipase is not very effective in the stomach, so, at most, 10% of lipids ingested will be broken down there. Nevertheless, anyone who habitually eats a high-fat diet will stimulate an increase in their lipase production.

The time it takes for a meal to pass through the stomach depends on the nature of the food eaten, but is usually between 2 and 6 hours. A meal that is very fatty will remain in the stomach for a long time and, if the stomach is distended, it may contribute to a prolonged sensation of fullness. On the other hand, a sugary drink will pass rapidly through the stomach, and digested glucose may appear in the bloodstream within 5 minutes of the drink being swallowed. But chyme does not leave the stomach in a continuous flow; it leaves in controlled bursts that depend on the motility of the stomach.

3.3.4 The small intestine

It is in the small intestine that most digestion and absorption takes place. If you look at Figure 3.7, you will see that there is a duct (small tube) which is connected to the small intestine almost immediately after its junction with the stomach. Secretions produced by the liver and the pancreas enter the gut via this duct – called the bile duct. The pancreas produces a rich mix of enzymes: a lipase, several protein-digesting enzymes and one starch-digesting enzyme (*pancreatic amylase*) which continues the digestion of polysaccharides. Other protein- and carbohydrate-digesting enzymes are secreted by cells in the walls of the small intestine.

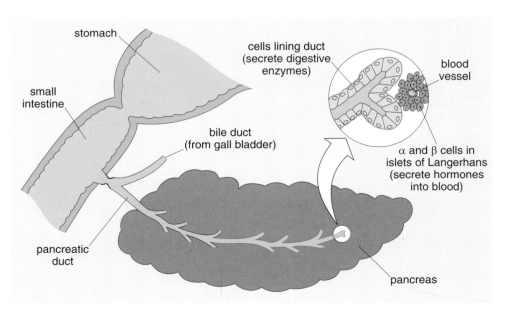

Figure 3.7 The bile duct and the pancreatic duct (not to scale), along which digestive secretions pass on their way to the gut. The pancreas also secretes hormones.

Both the pancreas and the liver produce bicarbonate that neutralises the acidity from the stomach. Additionally, the liver produces bile salts, which assist with the breakdown of lipids.

Lipids have to be digested in a different way from proteins and carbohydrates because they are not soluble in the *aqueous* (watery) solutions found in the gut. Because lipids do not dissolve in water, when mixed with an aqueous solution they separate out and form a layer on top of the liquid, in a similar manner to oil spilt at sea. Lipids associate together as globules or large droplets in aqueous solutions and the digestive enzyme lipase only has access to the outer molecules of the globule. If all lipid digestion occurred in this manner, as it does in the stomach, it would be a very slow and inefficient process. (Recall that it is necessary to break up large molecules into smaller units so that the enzymes in digestive juices have a large surface area to work on.) Therefore these globules need to be dispersed into much smaller droplets, and this occurs by the action of bile salts in a process known as **emulsification**. This is similar to what happens when washing-up liquid is added to greasy water. Emulsification greatly increases the surface area of fat droplets accessible to lipase (Figure 3.8). Lipase splits off two fatty acids from the triacylglycerols (TAGs). The result is a glycerol molecule with one fatty acid chain still attached to it (known as a *monoacylglycerol*) and two free fatty acids. It is the presence of fatty acids that stimulates endocrine cells of the small intestine to release the two anorexic hormones GLP-1 and PYY. As a fatty meal may have spent a long time in the stomach and little fat digestion occurs until the chyme is in the small intestine, it may be that GLP-1 and PYY cannot provide feedback to the brain sufficiently quickly after food is consumed to affect how much is eaten at that meal. This is an argument for eating slowly, maybe having a number of very small courses – like a Chinese banquet – and pausing between courses as a strategy for restraining food intake.

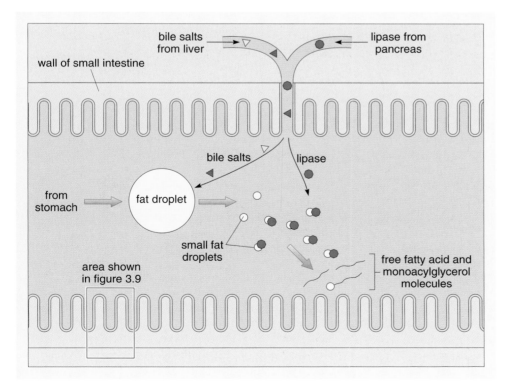

Figure 3.8 Summary of the emulsification and digestion of fats.

So the digestion of the macronutrients is completed in the small intestine and the products are available to cross the wall of the gut. The next section deals with this absorption.

3.4 Absorption and distribution of fuel molecules

The monosaccharides (e.g. glucose and fructose), amino acids and alcohol can all pass through the cells that make up the wall of the gut and find their way into tiny blood vessels within the gut wall (shown in Figure 3.9). Tiny blood vessels are known as **capillaries** and the areas where they are found, throughout the body, are known as *capillary beds*. A convention often used is that the capillaries that branch off from the larger vessels that have brought blood from the heart (**arteries**) are shown in red. The blue-coloured capillaries are shown feeding into larger vessels that return blood to the heart (**veins**). So, in general, arteries take blood to organs and veins collect blood back from organs. There is an interesting and slightly different arrangement in relation to the gut and liver. Blood from the capillary beds around the gut does not get taken straight back to the heart but instead goes to the liver in a vessel called the **hepatic portal vein** (Figure 3.10). Here there is another capillary bed.

The Greek *hepar-* means liver and the Latin *porta* means a gateway.

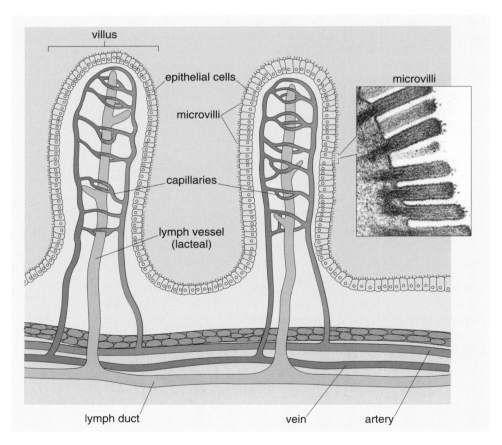

Figure 3.9 Diagrammatic cross-section showing the gut wall in the small intestine. Note how the villi (singular: villus) formed by the undulating wall of the gut greatly increase the area that is in contact with the gut contents. Each epithelial cell has microvilli, which further increase the area of contact.

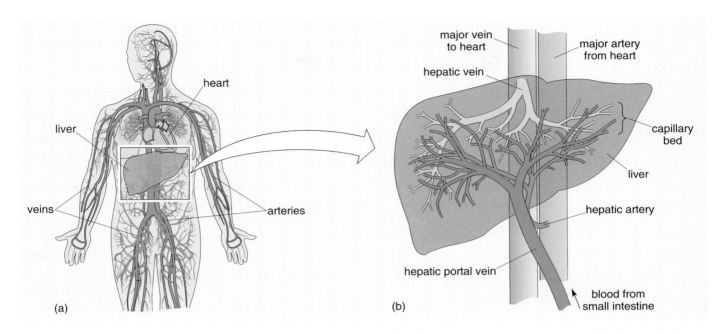

Figure 3.10 Simplified diagram showing (a) part of the vascular (blood circulatory) system and (b) the position of and blood supply to the liver. Note that blood flowing into the liver comes from both the hepatic artery and the hepatic portal vein. Blood leaves the liver through the hepatic vein.

The hepatic portal system is very important. It allows nutrients and other substances that have been absorbed from the gut to be withdrawn from the bloodstream by the liver. The liver is able to break down many toxic (poisonous) substances, including alcohol. In particular, it is vital to remove some of the glucose! High levels of glucose circulating in the bloodstream cause damage to cells, as you will learn in Chapter 4. In fact, if you were to take a Mars™ bar, 'digest' its carbohydrates into glucose and put that directly in your bloodstream, then your blood sugar level would be raised more than tenfold – to damaging levels. This emphasises just how carefully the processes for dealing with the ingestion, digestion, absorption and utilisation of food need to be regulated. In some respects, eating is a very dangerous enterprise, and the liver has a huge responsibility in ensuring that these potential dangers are minimised.

As you will see in Chapter 4, the liver cannot perform all these tasks instantly. You may be very aware, from your own experience or from the behaviour of others, that the liver is unable to break down immediately all the alcohol consumed when it arrives from the gut. Nor is the liver able to regulate the glucose level so that it stays at a *constant* value. It does, however, make very considerable adjustments to the blood glucose level while the blood is passing through the hepatic capillary beds. Once through the hepatic capillary beds, the blood, with its adjusted load of glucose, alcohol and amino acids, is returned to the heart and is then pumped out to the lungs, where it can pick up the oxygen needed by cells to metabolise completely the fuel molecules to release energy.

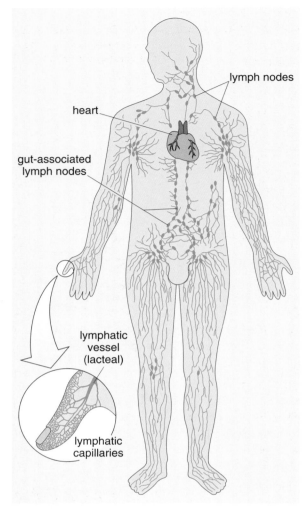

lymph nodes

heart

gut-associated
lymph nodes

lymphatic
vessel
(lacteal)

lymphatic
capillaries

Figure 3.11   The system of lymphatic vessels.
The inset of the thumb illustrates the extent of
the lymphatic capillary network. Note: the heart
is not actually part of the lymphatic system.

◆ What is (a) the name of the process that releases the energy
and (b) the name of the molecule that captures the energy
released from the fuel molecules (such as glucose)? (Hint:
Re-read Section 2.6 if unsure of the answer.)

◆ (a) The name of the process is cellular respiration. (b) ADP is
the molecule that captures the energy and makes ATP.

The dietary fatty acids and monoacylglycerols cannot gain
immediate access to the blood capillaries (as will be explained
later); instead, they are taken into the lymph ducts (Figure 3.9)
of the *lymphatic system* (Figure 3.11). This means that the liver
does not have the opportunity to adjust their levels as they leave
the gut.

From the inset in Figure 3.11, you can see that there is an
extensive network of vessels – the lymphatic capillaries.
These are closed-end vessels permeating the spaces between
cells. Fluid leaks out of blood capillaries into these *interstitial*
(between-cell) spaces and is then known as *interstitial fluid*.
It then drains into the lymphatic capillaries (where it is called
lymph), which return it to the bloodstream via a duct connecting
the lymphatic and *vascular* (blood circulatory) systems in the
neck. This one-way transport system has no pump.

◆ Which organ is the pump for the vascular system?

◆ The heart.

The lymphatic system depends on the squeezing and pumping
action of skeletal muscles to move fluid through its vessels.
This is why some obese people find that their extremities,
particularly their legs, have **oedema** (become swollen). As a
result of their inactivity, the return of the interstitial fluid to the
circulatory system is hampered. This is a particular problem
for legs because returning fluid to the heart involves working
against gravity. A further difficulty arises because the fat in the
abdominal cavity compresses the lymph vessels. Again, it is the lower body that is
worst affected, so swollen ankles tend to be an early symptom of oedema.

Depending on the size of the meal eaten and the mix of foods consumed, glucose
can be available in the blood within 5 minutes and may continue to be absorbed
over a period of hours. As indicated earlier, it is important to have glucose in the
blood to meet the metabolic needs (fulfil the energy requirements) of cells, but
levels must not rise to the point where glucose causes damage. Blood glucose
levels are maintained between approximately 3.5 and 6.5 mmol l^{-1}. Hormones
often control homeostatic processes and blood glucose level is maintained by
the pancreatic hormones **insulin** and **glucagon**. The hormone-releasing cells
of the pancreas are known as *islets of Langerhans* – shown diagrammatically
in Figure 3.7. Insulin is released from the β cells in the islets of Langerhans in
response to high blood glucose levels (usually following a meal) and glucagon

mmol l^{-1} (also seen as mmol/l) is
an abbreviation for millimoles per
litre. This is a unit of measurement
that is often used when measuring
substances within the blood. It
represents the amount of substance
per litre of blood.

is released from the α cells of the islets of Langerhans in response to low blood glucose levels. The action of insulin results in glucose being taken up by liver and muscle cells and being converted into glycogen (Figure 3.12). Glucagon activity reverses this process by initiating the breakdown of glycogen in liver cells and the subsequent release of glucose into the bloodstream. Note that insulin also encourages the uptake of glucose into adipose tissue, but here it is stored as fat and glucose cannot be released back into the blood from this source. The homeostatic maintenance of blood glucose levels will be discussed in more detail in Chapter 4.

α and β are the Greek letters alpha and beta, respectively, used here to refer to different types of cell.

Dietary fats do not appear in the bloodstream as rapidly as glucose. It may be an hour or two after eating before they are in the bloodstream. So eating fatty foods when your energy levels are low is not a good strategy. From Figure 3.12, you can see that many cells can use fatty acids instead of glucose to provide for their energy needs but, if not required for energy, adipose tissue and the liver will convert fatty acids to storage fat. It may be 10 hours after a meal before all the fatty acids provided by the meal have been cleared from the blood.

Amino acids travel from the gut to the liver in the hepatic portal vein, like glucose does. Dietary amino acids join other amino acids circulating in the blood. You should recall that these other amino acids have been derived from protein breakdown within cells (Section 2.4.3). This pool of amino acids is available as a resource for all body cells to enable them to synthesise proteins. The pool of amino acids in the blood does need to be topped up by dietary amino acids because protein is constantly being lost from the body. The liver can convert one type of amino acid to another to ensure that the pool has reasonable availability of all necessary amino acids. The diet must supply the eight amino acids that cannot be made by the liver (the essential amino acids), but many people consume more protein than is needed and the liver removes excess amino acids

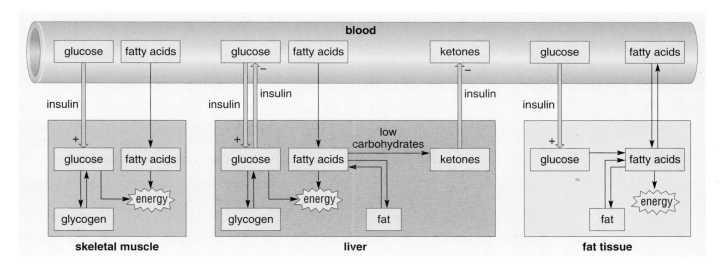

Figure 3.12   The relationship between glucose in the blood and its storage and release from muscle, liver and adipose tissue. Glucose can pass easily between the blood and the liver and back again. This is shown by the two-way arrow. Glucose can also pass easily into the muscle and adipose tissue, but in these cases glucose does not pass back into the blood. In muscle, glucose is converted into glycogen, and glucose released from glycogen is used to fuel muscle contraction. Adipose tissue converts glucose into fatty acids and thence to fat. When fat is broken down in normal metabolism, fatty acids are produced again, and may be released back into the blood and taken up by the liver. Here they are metabolised for energy and, if insufficient carbohydrate is present, ketones may be produced.

Excretion is the elimination from the body of a waste product that has been part of the body's fabric. Defecation is the elimination of waste that has never been part of the body's fabric; it has merely moved through the gut.

from circulation. These amino acids are processed to produce urea – a waste product that is *excreted* by the kidney – and fatty acids or glucose that will be used to provide energy or will be stored.

You have probably noticed that all the fuel molecules derived from macronutrients end up being stored as fat if they are consumed in excess of energy requirements. Alcohol cannot be converted into fat, so it cannot be stored as fat; instead, it is used to provide energy (ATP) in place of the other fuel molecules, so the outcome is the same. If you eat some bread and cheese with your glass of wine, the energy from the breakdown of the alcohol molecule will be captured by ADP and used to create ATP while the glucose, amino acids, fatty acids and glycerol from the digested bread and cheese will be stored as fat. This is assuming that total energy intake (TEI) has exceeded TEE and you have a positive energy balance.

This completes the overview of the absorption and distribution of the fuel molecules. We have, however, glossed over the way that molecules enter and leave cells. It is clearly not an entirely straightforward subject as there is no one mechanism suitable for all molecules. You may recall that the digested fatty acids and monoacylglycerols were not able to pass straight into the bloodstream from the gut, but had to be carried by the lymphatic system. It is the cell membrane that regulates entry into and exit from the cell. The membrane's selectivity is fine-tuned to the cell's needs and to its function, controlling the direction, amount and timing of the passage of chemicals. An understanding of the structure and function of cell membranes enables you to understand how hormonal and neuronal signalling works and how drugs – such as those designed to treat obesity – might interfere with such activity, and provides you with an explanation of how molecules pass across the intestinal wall. The next section concerns the passage of the products of digestion from the gut and will illustrate the general principles involved.

3.4.1 Movement across cell membranes

The cell membrane is responsible for maintaining the integrity of the cell. It keeps the internal fluids and *organelles*, such as the nucleus and mitochondria, separate from the external environment – the *extracellular* fluids. The membrane is very selective about what it will allow into the cell and also about what may leave. The consequence of this is that the *intracellular* fluid differs from the extracellular fluid. The concentrations of certain chemicals, such as glucose and oxygen, are different on the two sides of the membrane. In this situation, we say that there is a *concentration gradient* for each of these chemicals across the membrane.

The structure of the cell membrane is shown in Figure 3.13a. It has a homeostatic function in that it is responsible for protecting the finely balanced machinery inside the cell from changes in the extracellular fluid. At the same time, the cell's requirements for energy must be met and waste products must be eliminated. So the membrane must selectively permit the entry and exit of appropriate molecules.

Some very small substances can pass through the membrane freely, moving from the area where they are highly concentrated towards areas of lower concentration. (This process is called **diffusion**, and we say that the membrane is *permeable* to these substances. Oxygen enters cells by diffusion, as shown in Figure 3.13b.) Other substances are prevented from passing – for example, they might be too large.

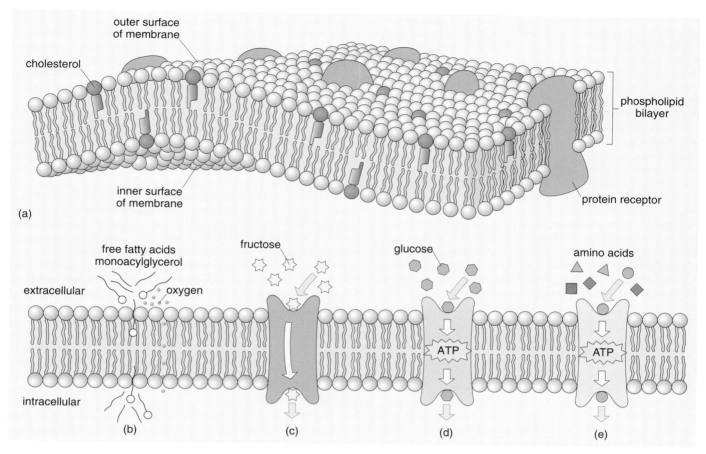

Figure 3.13 Transport across the cell membrane. (a) The structure of the cell membrane, with (b) oxygen, monoacylglycerol and fatty acids diffusing across. (c) The process of facilitated diffusion and (d, e) active transport. Note that both facilitated diffusion and active transport require the presence of specific protein receptors to be in the cell membrane and it is therefore the structure of the *particular* cell's membrane that determines how a substance may cross.

You can see that the major component of the cell membrane is a double layer (bilayer) of the fatty phospholipid molecules that were first described in Section 2.4.2. Although the design looks rather solid, fatty acids and monoacylglycerols can wriggle between the fatty phospholipids molecules, thereby diffusing across the cell membrane. So fatty acids and monoacylglycerol can get out of the gut and into the intestinal cells lining the gut by diffusion (Figure 3.13b). Not so the monosaccharides (e.g. glucose, fructose) or amino acids. They have to be helped to cross the membrane of the intestinal cells. (The molecules that help them to cross are membrane proteins.) Some amino acids and fructose cross by **facilitated diffusion** (Figure 3.13c). They cross from their area of high concentration in the gut to the area where they are at a lower concentration, inside the cell, by diffusion, but this is *facilitated* (helped) by membrane proteins changing their shapes so that they carry their 'cargo' across the membrane.

Finally, glucose and the majority of amino acids require energy to enable them to leave the gut. They have to 'lock onto' proteins that can be induced to actively receive them and 'carry them' across the membrane. This transporting mechanism involves the carrier protein changing shape and might be more appropriately described as 'kicking' the molecules across the membrane! This process is known as **active transport** because it requires an input of energy to drive it (Figure 3.13d, e).

◆ What is the name of the molecule that can provide the energy for active transport?

◆ ATP will provide energy for active transport.

You have now followed the absorption of the digested macronutrients from the gut into the intestinal cells lining the gut. The concentration gradients and the composition of the cell membrane on the opposite side, i.e. on the side nearest to the blood capillaries, is such that all the amino acids and the monosaccharides can pass into the bloodstream by facilitated diffusion. They will be carried by the hepatic portal vein to the liver. But something rather unfortunate has happened to the products of lipid digestion! The fatty acids and the monoacylglycerols have re-formed as TAGs, so now they are once again too big to wriggle across the membrane and leave the cell. Instead, the TAGs aggregate with cholesterol and phospholipids. A sac-like structure is formed that can be expelled from the intestinal cell and make its way into the lymphatic vessels (Figure 3.14).

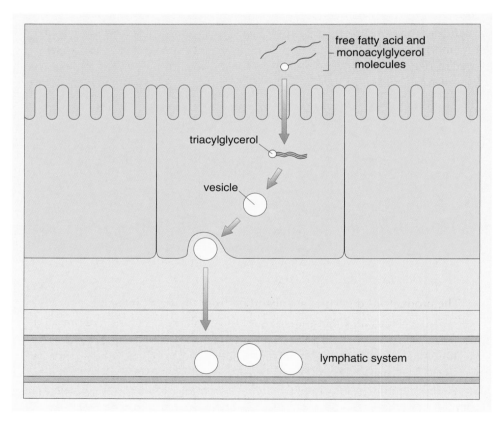

Figure 3.14 TAGs being synthesised in a gut cell and forming aggregates that are packaged into vesicles and exported into the lymphatic system.

You have now seen the methods used by the macronutrients to pass from the gut into the body's circulatory systems. Before leaving the cell membrane, we will say a word about the membrane proteins. In Figure 3.13a, the proteins appear to be fixed. In reality, there are many different types of membrane proteins with many different modes of action and they exhibit considerable fluidity in more than one respect. For example, there is a dynamic turnover of membrane proteins, and this is under genetic control (Section 2.4.3). Thus, as gene activity fluctuates, some proteins go but are replaced by similar proteins; others may be removed but not replaced.

◆ If a cell is actively making a protein, in what state is the gene said to be?

◆ A gene that is sending instructions for the manufacture of its protein is said to be *expressed.*

Newly made proteins may be immediately inserted into the membrane or they might be stored in the cytosol – for example, they could be stored within vesicles until required. When the density of a particular membrane protein is altered, it is said to be **downregulated** if numbers are decreased and **upregulated** if numbers are increased. By altering the availability of membrane proteins in this way, the cell can control which substances – and how much of any substance – can enter or leave the cell. Insulin enhances glucose uptake in muscle and adipose tissue by upregulating the number of carrier proteins available to transport glucose into the cell. In the absence of insulin, the receptors are downregulated and slip out of the membrane, back into the cytosol, where they are stored in vesicles until levels of insulin rise again. Insulin also enhances glucose uptake by the liver, but this is achieved by a different means that will be described in the next chapter.

3.5   Summary of Chapter 3

3.1   Macronutrients (fats, carbohydrates and proteins) must be broken down by digestion so that they can be absorbed into the body.

3.2   Enzymes – each one specific for one food type only – assist in the breakdown of the macromolecules.

3.3   Different foods take different amounts of time to pass through the gut. Glucose from a sugary drink will be in the bloodstream within minutes of being swallowed; most of a fatty meal may still be in the stomach a couple of hours after eating has finished.

3.4   The products of digestion are absorbed from the small intestine. Monosaccharides, amino acids and alcohol pass through the gut wall into the hepatic portal vein and from there to the liver. Monoacylglycerols and fatty acids pass into the lymphatic system and are transferred into the bloodstream in the neck region.

3.5   The liver regulates the level of glucose in the blood. The presence of the hormone insulin results in the liver taking glucose from the blood and converting it into glycogen – a storage product. Insulin also encourages glucose to be taken up by muscle cells and stored as glycogen, and to be taken up by adipose tissue and stored as fat.

3.6 Fatty acids circulating in the blood can be used as an energy source by most cells, but the excess will be stored in adipose tissue.

3.7 Excess amino acids are processed in the liver to produce a waste product, urea (which must be excreted by the kidney), and fatty acids and glucose, the excess of which will be stored as fat.

3.8 Alcohol can be used as an energy source; it cannot be stored (but the fuel molecules it replaces will be stored).

3.9 The structure of the cell membrane regulates the ways in which different molecules can enter and leave cells.

Learning outcomes for Chapter 3

LO 3.1 Define and use, or recognise definitions and applications of, each of the terms printed in **bold** in the text.

LO 3.2 Identify and explain the main parts and functions of the digestive tract and associated organs.

LO 3.3 Explain how fats, carbohydrates and proteins are digested and then absorbed from the gut.

LO 3.4 Explain how any of the macronutrients can contribute to body adiposity.

Self-assessment questions for Chapter 3

Question 3.1 (LOs 3.1 and 3.2)

In Section 3.1, it was revealed that digestion is mechanical as well as chemical. What are the mechanical methods that serve to break down macromolecules?

Question 3.2 (LOs 3.1, 3.2, 3.3, 3.4 and 2.3)

At the end of Section 3.1, it was explained that this chapter would help you to understand methods that might be used in the treatment of obesity. Suppose a drug was developed that was a lipase inhibitor. Would it be of assistance? Would it have any dangers associated with its use?

Question 3.3 (LO 3.4)

There are some groups of people who traditionally consume protein-rich diets (e.g. the men of the Masai tribe). Does this protect them from obesity?

CHAPTER 4

METABOLISM: THE BODY'S INTERNAL BALANCING ACT

The fire of life seems to burn brighter in some than others.

Dulloo and Miller (1987)

4.1 Introduction

There is a continuous need for food to provide us with metabolic fuels throughout life. In just one year, an adult human, maintaining their body weight, eats about 1000 kg of food! (It is difficult to imagine this quantity, but you should be familiar with the size of your own weekly food shop. Figure 4.1 shows a typical quantity of food that is eaten by a British family over the period of one week.) Our bodies routinely process the food consumed and distribute the products for either immediate use or storage; the control of the distribution of the macronutrients is the subject of this chapter.

In the previous chapter, you were introduced to how the body breaks down and uses all the different types of food that it takes in. You have become familiar with the various body organs involved in the digestive process and met some different enzymes, with specific jobs of breaking down different food types. Once the various macronutrients are broken down into smaller molecules, the body can absorb them and either use them or convert them for storage, saving them for later.

Figure 4.1 The typical quantity of food eaten each week by a British family of four, consisting of two adults, two growing children and their pets.

The main areas you will be covering in this chapter are how the products of carbohydrate and fat digestion are circulated and regulated throughout the body for use and, if they are not used, whether any excess is stored or discarded. You will then move on to consider the health implications of processing and storing those excess products of digestion, such as glucose and fatty acids. By the end of this chapter, you should have developed a better understanding of how the body deals with metabolic fuels during health and the development of overweight, obesity and disease.

4.2 Metabolism, homeostasis and a healthy balance

Metabolism is the collective name given to all the chemical reactions that take place within living cells; there are plenty of them! These reactions can be either *anabolic* (a building up or making of molecules) or *catabolic* (the breaking down of molecules, often involving the release of energy to be captured by ADP to make ATP). As mentioned in Chapter 3, these metabolic pathways involve a sequence of many small chemical reactions, each one requiring the presence of enzymes.

Anabolic and catabolic reactions are opposing processes. To bring some order to these opposing processes within the body, and to allow an organism to function properly, metabolism is regulated by homeostatic mechanisms. Metabolism is most efficient when each of the factors that contribute to the internal environment – the temperature, pH, nutrients, etc. – are present at optimum levels. For each of the factors, there is a small range above and below the optimum level that allows reasonably efficient metabolism, and in which homeostatic mechanisms are able to correct small deviations and return the level to optimum. However, when the level goes outside the tolerable limits for one factor, homeostatic mechanisms start to struggle. When this happens, there can be 'knock-on' effects to other systems, leading to the levels of other factors going outside their tolerable limits, resulting in a diseased state. In relation to obesity and many other metabolic disorders, blood glucose regulation is a key homeostatic mechanism.

pH is a scale used to measure the acidity of a substance.

Glucose is a major fuel source for most body tissues, and it is transported around the body in the bloodstream. You may think then that maintaining a high level of glucose in the blood would be a good thing. Unfortunately, **hyperglycaemia** (high levels of glucose in the blood), especially if present over prolonged periods of time, can cause undesirable effects in different tissues throughout the body. Although excess circulating glucose in the bloodstream can be removed and converted to fat for storage, prolonged exposure of tissues to high blood glucose levels (also called glucose toxicity) causes tissue inflammation and other damage. Such damage results from glucose molecules binding to protein molecules and causing irreversible structural changes, a process called *glycation*. The kinds of problems that may be experienced as a consequence include damage to the eye, to red blood cells, to nerves in body extremities and to kidneys – and an increased risk of cardiovascular diseases. High glucose levels in blood also provide a nutrient-rich environment for pathogens such as bacteria, leading to an increased susceptibility to infection. Very high glucose levels in the blood pulls water out of cells, which greatly impedes metabolism and can be fatal. For all of these reasons and more, a sustained high level of glucose in the blood is a bad thing.

Pathogens are harmful micro-organisms such as bacteria, viruses or fungi that cause disease.

Equally, a low level of glucose in the blood is also undesirable. Brain cells need a continuous supply of glucose, and if levels of blood glucose fall below the optimum range, these cells will not be functioning very well, and will be unable to carry out their basic metabolic processes. It follows that the level of glucose in the blood should neither be too high or too low; there is an optimum range. The mechanism that maintains glucose within the optimum range is homeostasis.

Blood glucose control will be looked at in detail in this chapter to help you understand its importance in the maintenance of health and also in the development of overweight, obesity and disease.

4.3   Insulin and the regulation of blood glucose levels

Glucose is used by metabolising cells, where it is broken down to carbon dioxide and water by the chemical process of cellular respiration (Figure 2.6). In respiration, energy is released and captured in the form of ATP. But before the cells within the body can metabolise glucose, the glucose has to get to them and then into them. Let's remind ourselves about how glucose gets into the bloodstream. A food containing carbohydrates or a drink containing glucose is ingested. Before absorption, the larger carbohydrates need to be digested into smaller units, including glucose, within the gut, but the glucose from the drink is available much more rapidly for absorption into the bloodstream.

The bloodstream transports glucose to all cells, but you will recall from Section 3.4.1 that glucose cannot just drift across the cell membrane into the cells. Facilitated diffusion enables glucose to get from the bloodstream into metabolising cells. For this to happen, two conditions must be met. Firstly, there must be a higher concentration of glucose in the blood than in the cell. The second condition that must be met is that dedicated carrier molecules must be present in the cell membrane. These molecules, are membrane proteins (Section 3.4.1) called *glucose transporters* that span the cell membrane and assist the movement of glucose across it.

◆ Cells that are working hard have high metabolic demands and are therefore using ATP continuously, so they rapidly use up their glucose. How does this affect facilitated diffusion?

◆ The cell that uses up its glucose will have a very low concentration of glucose. Hence the concentration gradient across the membrane from the bloodstream to the cell will be steep, making it easier for glucose to cross the membrane into the cell.

◆ Name one organ with high metabolic demands. (You are currently using it!)

◆ Your brain, although you could have mentioned other important organs, such as the liver, lungs or muscle.

Let's consider the importance of glucose for the brain and the nervous system. The supply of glucose to the brain is essential, and accordingly its uptake is unrestrained. The brain can take up glucose regardless of the insulin status of the body because it has glucose transporters that are not regulated by insulin. This is not the case for most other body tissues. Muscle and adipose tissue, for example, have transporters that are regulated by insulin. In these tissues, cells keep their glucose transporters out of the way, packaged into vesicles in the watery cytosol of the cell until insulin signals that the transporters are needed in the cell membrane. These transporters are then moved to the cell membrane and are inserted into it (Figure 4.2). This is known as upregulating the transporter (Chapter 3).

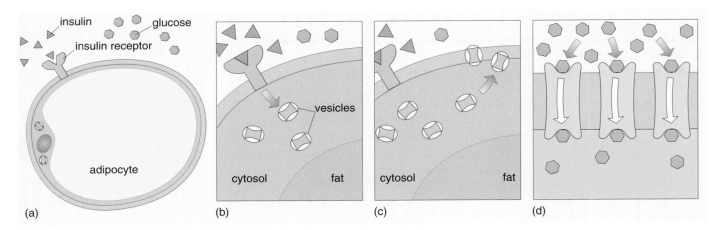

Figure 4.2 The upregulation of glucose transporters in an adipocyte. (a) Insulin and glucose from the bloodstream are present near an adipocyte. (b) Insulin locks onto the insulin receptor embedded in the cell membrane which triggers the insulin receptor to send a message via the cytosol to the vesicles holding the glucose transporter. (c) The vesicles respond by moving towards the cell membrane. (d) The vesicles release the glucose transporters, which insert into the cell membrane and facilitate the movement of glucose into the cell.

◆ When insulin levels are very low, which cells will be able to draw glucose from the bloodstream?

◆ Brain cells, because insulin is not required by the brain to obtain glucose from the blood.

Because the brain must keep working and depends on glucose as its main energy source, the body has mechanisms to ensure the supply of glucose during times of need. The times of need are the periods between meals, after the previous meal has been digested and nutrients absorbed, and before the next meal is taken – known as **fasting**. The severity of the need often determines the scale of the response. Under fasting conditions, ranging from an **acute** (short-term) overnight fast, glucose uptake by most of the rest of the body excluding the brain is restricted by the lack of insulin, thus protecting its preferential delivery to the brain. But if food is in short supply, as in **chronic** (long-term) famine conditions, circulating blood glucose from any small meal is likely to be used up rapidly. It is not too surprising that there are a few other energy supply options available. As you read in Chapter 3, the liver stores glucose by converting it into glycogen. Subsequently, the glycogen can be broken down to provide enough glucose for other tissues to be maintained for a few hours. Before this glycogen reserve is completely used, glucose starts to be made from other molecules – mainly by the liver, but also by kidney cells – by the process known as **gluconeogenesis**. This is how the body avoids blood glucose levels from going too low – during short-term

You break your overnight fast with your first meal – hence breakfast.

Gluco is short for glucose, neo means new and genesis means make.

fasting, it can produce its own glucose from other available non-sugar precursors. In some extreme diets and more extreme conditions of starvation, another fuel source is produced that can be used by the brain. (More about these conditions later, after we have covered the normal 'fed' situation.)

We have just established the importance of blood glucose as an energy supply to the brain and how glucose uptake by the brain and the nervous system is not influenced by insulin. In a number of other tissues, insulin does control their access to blood glucose – namely the liver, skeletal muscles and adipose tissues. The hormone insulin is well known for its actions controlling blood glucose levels, partly because a complete lack of insulin quickly results in coma and death. You may know of individuals with diabetes mellitus who are unable to produce insulin and who need to inject themselves daily with replacement insulin. (Diabetes is considered in Chapter 7.) We will now look more closely at the various roles of insulin in normal blood glucose control.

Glucose levels in the bloodstream are tightly regulated and maintained within a narrow range, as you will see in the next section. In Chapter 2, body temperature was used as an example to illustrate a homeostatic system. Here we are considering a similar representation of such a system but with different circuit components (Figure 4.3). The two main circuit components are:

1   pancreatic cells that secrete the hormone insulin (in response to increasing blood glucose levels)

2   body tissues that respond to insulin (mainly skeletal muscles and adipose tissue) by taking up glucose from blood into their cells, and the liver, which is stimulated to store glucose as glycogen.

As you will see, a change in any part of the circuit will cause effects elsewhere within the system. This circuit is the foundation for understanding what is at the core of many metabolic disturbances involving blood glucose control. Further detail will be gradually added in the following sections to help you understand this complex system.

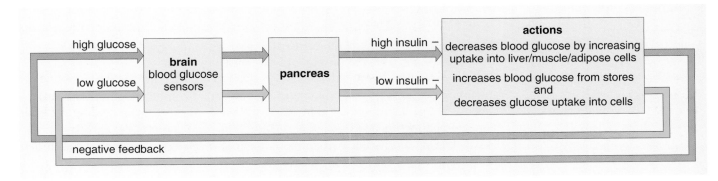

Figure 4.3   Simple representation of the control circuit for regulating blood glucose levels. Insulin is produced from pancreatic cells in response to high glucose in the blood. Insulin acts on target cells of liver, skeletal muscle and adipose tissue, causing them to take up glucose from the blood. Later, when sensors in the brain detect that blood glucose levels have dropped below the optimum value, the pancreas reduces its production of insulin and cells decrease their uptake of glucose. At this point, blood glucose is released from storage or resupplied from food, to increase blood glucose levels and so the process – an example of negative feedback – continues.

Summary

Blood glucose levels are tightly regulated by a homeostatic mechanism involving the hormone insulin. Glucose can enter the brain and the nervous system *without* the help of insulin. It can only enter liver, skeletal muscle and adipose tissue cells *with the help* of insulin.

Prolonged high blood glucose levels are toxic to some tissues; very low blood glucose levels can lead to coma and death. Once the supply of glucose from a meal drops to a low level, the liver and kidney can replenish supplies by breaking down glycogen stores. Before these stores are exhausted, the liver begins to manufacture glucose from non-carbohydrate precursors.

4.3.1   After feeding: the role of insulin

Insulin is widely associated with blood glucose because of its fundamental role in controlling glucose uptake into cells. Insulin also has other functions within the body, some of which are related to blood glucose control and some of which are not, including influencing cell growth and development. Many hormones have multiple and sometimes unrelated functions within the body. In the normal, healthy state, the control of blood glucose levels depends on the efficient action of the hormone insulin. In relation to developing overweight and obesity, we are mainly interested in insulin's effects on blood glucose control, covered below, and also insulin's effects on fatty acid metabolism, covered later in Section 4.4.

◆ Gluconeogenesis is a process that is influenced by insulin. Insulin is secreted when blood glucose levels are high, so do you think insulin will promote or inhibit gluconeogenesis? (Hint: Think about whether more glucose is required in the blood.)

◆ Insulin will *inhibit* gluconeogenesis: when blood glucose levels are high, there is no requirement for more glucose.

◆ What effect do you think insulin will have on the conversion of glucose to glycogen (**glycogenesis**)?

◆ Glycogen is a storage molecule produced when glucose levels are high. So insulin is likely to promote glycogenesis.

Insulin has a number of complementary actions that combine to maintain blood glucose levels within the optimum range. It slows gluconeogenesis in the liver. Gluconeogenesis makes new glucose from other non-sugar molecules (see Box 4.1 later). It is needed when blood glucose levels fall outside the optimum range (e.g. when there is a long time between meals). Insulin is circulating when blood glucose levels are high, so it makes sense for insulin to switch off gluconeogenesis.

Insulin increases the conversion of glucose to glycogen (glycogenesis) in liver and muscle cells by stimulating the relevant enzymes. It also inhibits the release of glucose from glycogen (**glycogenolysis**) in liver and muscle cells by its inhibitory effects on the enzymes for this metabolic pathway within the cells.

Once glycogen stores are replenished (i.e. full), excess glucose is converted to another molecule called acetyl CoA. Acetyl CoA can be used for fatty acid synthesis (see Figures 2.6 and 3.12, and also Section 4.4).

Look at Figure 4.4a, which shows the blood glucose levels of a person over 24 hours.

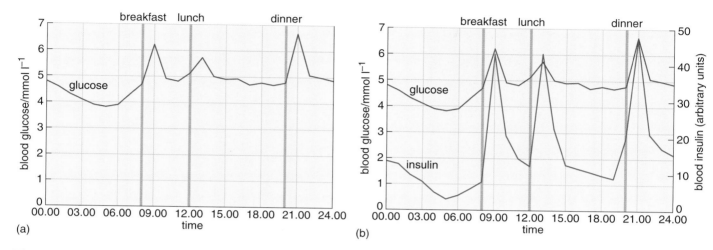

Figure 4.4   (a) Glucose levels and (b) glucose and insulin levels in the blood of a healthy individual over a 24-hour period; three meals were eaten and they slept overnight (from 23:00 to 07:00). Meal periods are shown as darker vertical bars.

◈   Using the left-hand axis, the blood glucose scale, what are the lowest and the highest levels measured? When do peaks occur?

◆   The lowest level is just under 4 mmol l^{-1}. The highest level is over 6 but less than 7, and is about 6.6 or 6.7 mmol l^{-1}. The peaks occur about one hour after meals.

◈   What is the range of glucose levels during sleep?

◆   During sleep, blood glucose is maintained at between about 3.7 and 5 mmol l^{-1}.

Two features of glucose levels are particularly striking. Firstly, the rapid rise and fall of glucose level after a meal. Secondly, the near-constancy of blood glucose levels between meals. Both these features indicate that blood glucose level is closely regulated in the 'normal' state.

Now look at Figure 4.4b. The purple line still represents glucose levels. The additional green line represents insulin levels. (Note the scale for the insulin is shown on the right-hand vertical axis. The actual units in which the insulin level has been measured are very small and need not concern you; we have called them arbitrary units for simplicity.)

◈   Look at Figure 4.4b, which shows the level of insulin in the blood. How do insulin levels change in relation to the meals?

◆   Insulin levels rise dramatically after meals, peak and then fall rapidly.

1988
coffee
(with whole milk and sugar)
45 calories
250 ml

2008
mocha coffee
(with steamed whole milk
and mocha syrup)
350 calories
500 ml

energy content difference: 305 calories

Figure 4.5   An example of 'portion distortion'. This term describes how a food item has changed in both quantity and energy content over time, often without the awareness of the consumer. The volume of drink has doubled and the energy content has increased to almost 8 times the original amount.

The graph shows very clearly that glucose and insulin levels rise and fall together. As the glucose level rises, so does the level of insulin. Insulin encourages the uptake of glucose into cells. We will return to consider the rise in insulin before dinner in Chapter 5. The rise of both insulin and glucose levels before breakfast is a consequence of fasting and the release of glucose from the liver.

◆ From Figure 4.4b, can you say that glucose causes the release of insulin?

◆ The scale on the graph is such that it is difficult to say whether levels of glucose rise before the levels of insulin rise or vice versa. Either way, these data only present a correlation; they could not prove a causal relationship.

Now look at Figure 4.5, which shows an example of how some common convenience or 'takeaway' foods have changed in terms of quantity and energy content over time. This is sometimes called 'portion distortion'.

◆ Using Figure 4.4b, describe the effect you think each drink of coffee will have on blood glucose and insulin levels. (Note: both sugar and milk contain glucose.)

◆ Both drinks of coffee will cause blood glucose and insulin levels to rise. As there is double the volume of coffee and approximately eight times the energy content in the 'modern' coffee, you could expect there to be an appropriate increase in glucose and insulin within the blood as the contents of this coffee are digested and absorbed.

You might want to think about this over your next portion of coffee!

Later in this chapter and in Chapter 7 you will learn about the effect of raised blood glucose and insulin on health and the development of overweight and related disease over time.

Summary

The level of glucose in the blood is maintained within certain limits by homeostatic mechanisms. One component of the homeostatic mechanism is the hormone insulin, which has a number of complementary actions. Insulin stimulates glycogenesis (and conversely inhibits glycogenolysis) and slows gluconeogenesis. When no more glycogenesis can take place because the liver, kidney and muscles' storage capacity is reached, insulin stimulates the conversion of glucose to fatty acids via an intermediary molecule called acetyl CoA.

4.3.2   Feedback loops

Let's now develop the process in a little more detail and introduce another component of the glucose homeostatic mechanism – glucagon. Glucagon is a hormone that, like insulin, is produced by the pancreas.

◆ What are the specialised endocrine cells within the pancreas that secrete the hormones insulin and glucagon?

◆ The specialised cells within the pancreas are called islets of Langerhans (Figure 3.7).

Before going any further, please be aware that in this chapter you will repeatedly meet the hormone called glucagon and the storage form of glucose in the liver and muscles called glycogen. It would be easy to be confused if you are not clear about the functions of these very different but similarly named substances.

Returning to insulin and glucagon, these are hormones involved in controlling blood glucose levels; they have opposing actions and, by working together, they achieve a regulated blood glucose level. Release of insulin into the bloodstream is stimulated by food, its smell, its presence in the stomach and also by high blood glucose levels. Blood glucose levels of between 6 and 7 mmol l^{-1} are reached quickly after a person has eaten a snack or meal and starts digesting and absorbing their food. This means that **postprandial** ('following a meal') blood glucose levels are much higher than those in tissue cells. One action of insulin is to promote the uptake of glucose into cells of liver, muscle and adipose tissue and thereby reduce blood glucose levels.

Glucagon is a hormone with opposing actions to insulin. Glycogen is the storage form of glucose.

Activity 4.1   Interpreting flow diagrams

Allow 10–20 minutes for this activity

Figure 4.6 is a flow diagram. You can read this and other flow diagrams just like text: 'high blood glucose leads to the pancreas increasing insulin secretion', and so on.

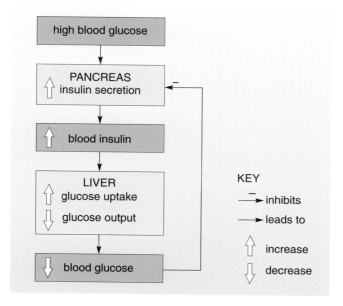

Figure 4.6   Flow diagram showing the role of insulin in the regulation of high blood glucose levels and the involvement of the liver, one of the target tissues of insulin action.

◆ If you look at the top of the figure, does the flow diagram begin just after a meal or during a period of fasting?

◆ The high blood glucose at the top of this diagram suggests that there has been a recent intake of food or drink containing glucose that has then been digested and absorbed.

This input to the system has disturbed the homeostatic equilibrium.

Start from the top of Figure 4.6 and, using the key, work your way through the diagram, stopping when you get to pancreas a second time. Note that you finished by saying 'a decrease in blood glucose inhibits insulin secretion from the pancreas'. It is this inhibition that defines this as a *negative feedback* system; the minus sign represents the inhibition, and is feedback because a later event in the sequence influences the beginning of the next cycle through the system.

Now, using Figure 4.7, which has low blood glucose at the top (following the decrease in blood glucose at the bottom of Figure 4.6), write a short paragraph explaining what happens to the secretion of insulin and its subsequent effects in the liver when blood glucose level is low. Complete the flow diagram as you go. Stop when you have returned to the bottom again.

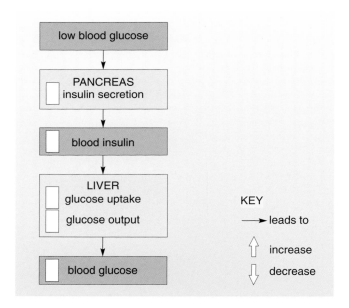

Figure 4.7 The regulation of low blood glucose levels and the involvement of the liver, one of the target tissues of insulin action. Complete this flow diagram by adding arrows to the white boxes.

Now look at the comments on this activity at the end of this book.

We have already established that following a meal glucose levels are higher in blood than in tissue cells, but there are some cells that glucose can only enter via transport molecules located in their membranes. While eating and digestion is taking place, insulin enters the bloodstream, and promotes entry of glucose into muscle and adipose cells by upregulating the number of special glucose transporter molecules in their membranes. Insulin has a *different* effect on liver cells, stimulating the enzyme *glycogen synthase* to convert glucose to glycogen (Figure 4.8), but the outcome is the same: the liver cells take glucose out of the bloodstream.

The synthesis of glycogen within liver cells effectively reduces the concentration of glucose inside these cells.

◆ How does the lower concentration of glucose within the liver cells help glucose get into the cells?

◆ The low glucose concentration inside the cell (because it is being converted to glycogen) maintains a concentration gradient across the cell membrane with high blood glucose concentrations outside and low glucose concentrations inside the cell. This promotes further entry of glucose into the cell.

We have discussed at length the first action of insulin, enabling the uptake of glucose by cells. The second action of insulin is to prevent the release of glucose from glycogen stores. Insulin does this by stopping the first stage of glycogenolysis by preventing the key enzyme, *glycogen phosphorylase*, from working (Figure 4.8). So when glucose is in the bloodstream, insulin is also present in the bloodstream and glycogen stores are maintained.

A third action of insulin is to stimulate storage. Insulin does this in three distinct ways, illustrated in Figures 4.8 and 4.9. Firstly, as mentioned above, by activating enzymes that convert glucose to glycogen. Secondly, also as mentioned previously, by stimulating the conversion of acetyl CoA from glucose to form fatty acids (Figure 4.9a). Thirdly, glucose can be converted to glycerol, which can combine with free fatty acids to make triacylglycerol (TAG; Section 2.4.2), which is of particular relevance to weight gain. This synthesis of TAG occurs in adipocytes (Figure 4.9b). So while insulin is at high enough concentrations in the blood and glucose and free fatty acids are available, as they would be after a meal, TAGs will be made and stored in adipose depots throughout the body. In terms of weight gain, if insulin levels are kept high and the appropriate precursors (building blocks) are readily available, fats will be continuously made and stored, ultimately leading to excess weight gain. We will return to this point again later.

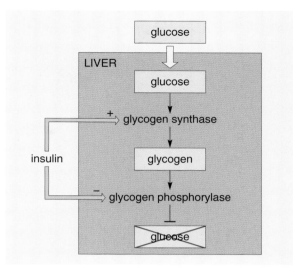

Figure 4.8 Schematic diagram showing the influence of insulin on glycogenesis and glycogenolysis in the liver. (The plus sign (+) indicates a stimulatory effect.)

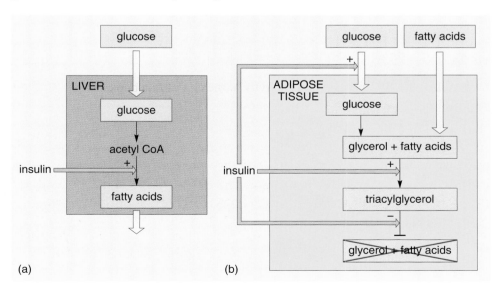

(a)

(b)

Figure 4.9 (a) Conversion of glucose to fatty acids via acetyl CoA in the liver under the influence of insulin. (b) Formation of TAG from fatty acids and glycerol in adipose tissue under the influence of insulin.

As a consumed meal is digested and the products absorbed and distributed about the body, insulin levels fall as blood glucose levels decline, normally between 1 and 2 hours after a meal (Figure 4.4b). The combined effect of low insulin and low glucose levels results in the secretion of glucagon by the pancreas, so to your list of the actions of insulin can be added a fourth: high levels of insulin inhibit glucagon secretion.

The effects of glucagon on target cells are the opposite to those of insulin, in that they promote increases in the levels of blood glucose. The major role of glucagon is to maintain levels of blood glucose during the **postabsorptive phase**, when digestion has finished and the transport of blood glucose derived from digestion to target cells is complete, and is the subject of the next section.

Summary

Glucose, having entered a cell, will either be broken down to release energy or be converted into another molecule for storage. Glucose is stored in the muscles and liver as glycogen or in adipose cells as TAGs. Conversely, when insulin levels are low, fewer glucose transporter molecules are available and less glucose will enter target cells, for either use or storage.

4.3.3 Fasting: the role of glucagon

During fasting, if blood glucose levels were allowed to continue to decline, the brain would not have access to sufficient glucose for metabolism. As you have already seen, glucose is the molecule that the brain normally uses to provide energy. You may have experienced feelings of dizziness or 'light-headedness' when you are very hungry. These effects are caused by a low blood glucose level of less than 3 mmol l^{-1}: a condition known as **hypoglycaemia**. Situations that can lead to a drop in blood glucose levels include intensive prolonged exercise and extreme cases of starvation. Extreme hypoglycaemia can lead to coma and eventually death.

During short-term fasting, glucagon prevents blood glucose levels from dropping too low by stimulating the breakdown of glycogen stores in the liver and elsewhere (Figure 4.10).

◆ The previous sentence contains both the terms glucagon and glycogen. Without looking back, which one is the hormone?

◆ Check your answer with Section 4.3.2.

The enzyme that enables glycogen breakdown is glycogen phosphorylase. (You will recall that this was the enzyme mentioned earlier that is prevented from working by insulin.) This enzyme is activated by glucagon. A second action of glucagon is to stop the conversion of glucose into glycogen, and hence stop glycogen stores being replenished. Glucagon does this by preventing the enzyme glycogen synthase from working. (Again, this is the opposite effect to insulin.) Both these effects occur in the liver and are shown in Figure 4.10.

A third action of glucagon is to enhance the breakdown of TAGs to glycerol and fatty acids, for further use as an additional energy source (Figure 4.11).

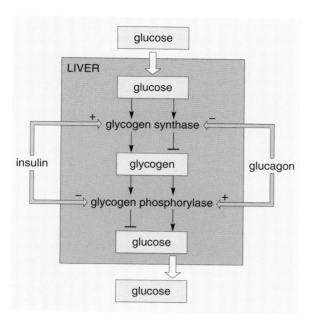

Figure 4.10   Interconversions between glucose and glycogen within the liver that are influenced by insulin and glucagon.

A fourth action of glucagon is to activate the enzymes involved in gluconeogenesis (see Box 4.1) so the liver returns to making glucose as none is being supplied via ingested food. (These are not the only actions of glucagon, but they are sufficient to demonstrate glucagon's key role in glucose homeostasis.) And so a cycle exists between insulin and glucagon: insulin rises and glucagon falls with the arrival of glucose from food, influenced by how much glucose has to be dealt with; subsequently, insulin falls and glucagon rises during fasting.

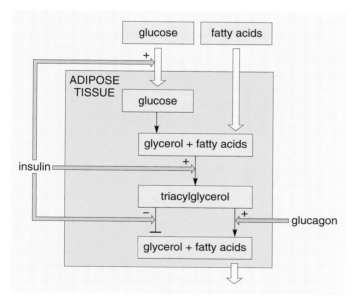

Figure 4.11   The effects of insulin and glucagon, inhibiting and stimulating the breakdown of TAGs, respectively.

Box 4.1   Alternative fuels and ketones

The pathways for glucose, fatty acid and amino acid metabolism in humans share some common intermediates. Just as footpaths meet and cross, allowing the same destination to be reached from different starting points, so common intermediates allow some but not all of these metabolic fuels to become each other, i.e. to be interconverted (Figure 4.12).

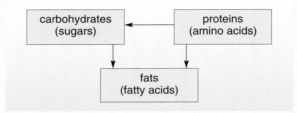

Figure 4.12   Interconversions between the three main metabolic fuels.

It is worth briefly expanding on the cell's metabolic response when glucose is not in excess, but in short supply. Such a situation can occur in the fasted state, during vigorous exercise, as a result of starvation and also in people who lack insulin – a condition known as diabetes mellitus (Chapter 7). Essentially, there are mechanisms that release glucose from storage, as already discussed, or make use of alternative energy sources.

The alternative energy sources are of two types. One type of energy source produces glucose from non-carbohydrate sources. Most amino acids (from proteins) and glycerol (available after TAG breakdown; Figure 2.6) can be used for gluconeogenesis. Note that glucose can be synthesised only from the glycerol portion of TAG molecules, not from fatty acids. The other molecule that can be used for gluconeogenesis is *lactate*. Lactate is produced in skeletal muscles from glucose during vigorous exercise. When exercising vigorously, glucose is rapidly broken down to pyruvate in muscle cells, providing ATP to power the muscle contractions. As you can see in Figure 2.6, the next stage in the production of ATP involves the pyruvate molecule entering the mitochondrion *and* it requires oxygen to be available. In vigorous exercise you get 'out of breath' and are unable to provide an adequate supply of oxygen, so pyruvate is converted to lactate, which does not require oxygen but does supply a little more energy to help keep the muscles working.

The fatty acids are the second type of energy source. Fatty acids can be metabolised to yield energy and indeed provide more energy per gram than either carbohydrates or proteins (Chapter 2). The initial stage of fatty acid breakdown produces acetyl CoA (Figure 2.6) and energy by chopping small pieces from the end of the fatty acid chains. Each fatty acid produces a number of acetyl CoA molecules; the longer the chain, the more acetyl CoA and the more energy is derived from this initial stage (Section 2.6). When glucose is severely depleted, it affects the efficiency of the TCA cycle such that the TCA cycle cannot accept all the acetyl CoA. The excess of acetyl CoA enters an alternative metabolic pathway in the liver cells and produces molecules called *ketone bodies* (or ketones). Hence ketones are only formed when there is an excess of acetyl CoA available. Ketone bodies are released from the liver to the blood circulation as the liver does not use them. They are used by other tissues as a fuel source; in the case of the brain, they can cross the *blood–brain barrier*, the protective arrangement of cells that prevents many substances from reaching the brain. Unfortunately, at high concentrations ketone bodies are toxic because they increase the acidity of the blood to unacceptable levels, a situation known as *acidosis* (i.e. homeostasis of the blood's pH is not maintained). If acidosis persists, death results, in hours or, at most, a few days.

So far, we have examined the postprandial control of blood glucose levels by the hormones insulin and glucagon. We have also touched on the subject of energy storage when glucose levels are high, and alternative energy sources when glucose levels are low. By now, you should also be starting to appreciate that there is 'cross-talk' between metabolic pathways and body systems. The links between blood glucose status and the maintenance of adipose depots throughout the body are considered in the next section, at the end of which these metabolic systems will be drawn together in one interconnecting diagram.

4.4 Creating an energy reserve: longer-term storage and use of lipids

The *storage* of fat in adipose tissue as TAGs follows fatty acid synthesis in the liver, whereas the *use* of these stores involves fatty acid breakdown. These are coordinated events within tissues, finely tuned so that the conditions that promote fatty acid synthesis work against fatty acid breakdown. The reverse is also true; conditions that promote fatty acid breakdown work against fatty acid synthesis. These metabolic pathways are regulated by enzymes for the specific steps involved in the reactions and are also regulated by hormones. You will not be too surprised to learn that the key hormones involved here are insulin and glucagon. This is a major example of the cross-talk mentioned above between regulatory systems within the body.

◆ What is meant by 'cross-talk between regulatory systems'?

◆ The same regulatory factors are involved in different regulatory systems – for example, the hormones insulin and glucagon in glucose homeostasis and the maintenance of adipose depots.

In well-nourished individuals, insulin activates fatty acid synthesis in the liver and TAG synthesis in adipose tissue (Figure 4.9). In addition, insulin inhibits **lipolysis**, the breakdown of TAGs. These effects of insulin serve to decrease the general availability of fatty acids. So under the influence of insulin, fatty acids are made in the liver and (after release into the bloodstream) are taken up by adipose tissue, where they combine with glycerol to make TAGs. The TAGs are stored, and adipose tissue reserves increase in size. Hence when there is excess glucose available, the body protects itself from hyperglycaemia and its associated health consequences (Section 4.5 and Chapter 7) by removing it from the circulation and after various biochemical conversions stores it away as fat.

The opposite situation is under the influence of glucagon, which mobilises fuel in times of need. Glucagon inhibits fatty acid synthesis in the liver. Glucagon also activates another pathway that leads to the breakdown of TAGs – that is, it stimulates lipolysis in adipose tissue, increasing the availability of fatty acids by the breakdown of TAGs stored there. The released fatty acids are transported and used elsewhere (Figure 4.11).

Up until now in this chapter, we have mainly considered the regulation of blood glucose levels in a healthy body. We will now move on to what happens within the body when certain aspects of the system are out of balance, which leads to the development of unhealthy physiological states and, ultimately, disease.

4.5   Insulin resistance

The worldwide rise in overweight and obesity appears to have been accompanied by a rise in diabetes, which itself is thought to develop after a period of insulin resistance. **Insulin resistance** develops when target tissues such as the liver and muscle begin to lose their ability to respond appropriately to insulin, reducing the ability of those tissues to take up glucose. As a consequence, more insulin is produced, so the level of insulin in the blood is high, a condition known as **hyperinsulinaemia**.

◆ Why do you think the condition results in high insulin levels?

◆ As the tissues are becoming more resistant to the actions of insulin, the brain continues to detect too much glucose in the circulation. The normal response to too much glucose is to stimulate the pancreas to produce more insulin. Hence there are high circulating concentrations of insulin.

However, in the initial stages of insulin resistance, hyperinsulinaemia is accompanied by normal glucose concentrations.

◆ Why do you think blood glucose concentrations are normal despite insulin resistance?

◆ Insulin resistance *reduces* the ability of cells to respond to insulin. The cells are able to respond to high levels of insulin in the initial stages.

You may have also mentioned that the brain cells can take glucose out of the bloodstream without the need for insulin. In fact, several tissues in addition to the brain – including the nerves of the arms and legs, the lens and other tissues in the eye, the kidneys, small blood vessels and red blood cells – manage glucose uptake independently of insulin. Such cells do not become insulin resistant and continue to use glucose and remove it from the bloodstream.

Over time, as insulin resistance progresses, glucose struggles to get into the insulin-resistant target tissues. The target tissues are therefore deprived of glucose while the blood glucose concentrations rise and hyperglycaemia results. So despite the body being in a well-fed state, the liver responds as if it is fasting: gluconeogenesis is stimulated, which releases yet more glucose into the bloodstream. Similarly, in adipose tissue, fat reserves may be broken down, releasing fatty acids and glycerol into the bloodstream so that the liver can make yet more glucose. However, because insulin is present, normal fatty acid metabolism can also take place. This means that, as long as the precursors (glycerol and fatty acids) are readily available – which they would be in the diet of someone who was well fed – TAG synthesis will occur, causing adipose depots to expand.

The pancreas, though, cannot keep producing large quantities of insulin and will eventually reduce its production. The level of insulin in the blood will fall, followed by an increase in blood glucose. At this point, the individual may lose some body weight as the insulin-resistant body continues to act as if it is fasting and mobilises body reserves. With disease progression, medication in the form of insulin is now required. You will realise from your knowledge so far that this is likely to result in further weight gain (Figure 4.13).

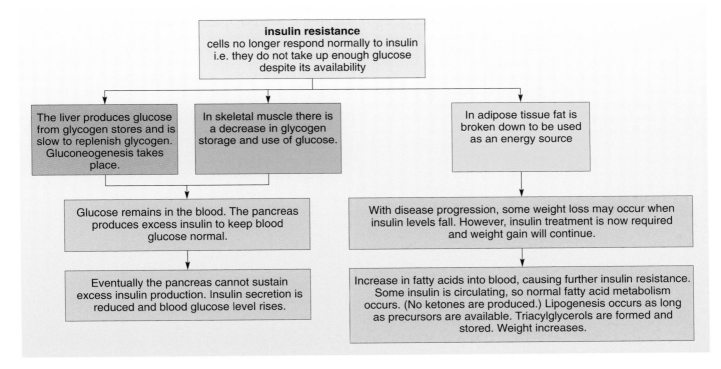

Figure 4.13   The progression of insulin resistance.

Individuals of normal weight may have insulin resistance. However, the situation is exacerbated (increases in severity) in overweight and obese individuals because the presence of excess circulating fatty acids in the bloodstream causes insulin resistance in target tissues.

As mentioned above, several tissues in addition to the brain manage glucose uptake independently of insulin. Hence, under conditions of high blood glucose, there will be a steep concentration gradient, and cells in these tissues will tend to be subjected to too great an influx of glucose. These tissues are very susceptible to becoming damaged as a consequence of glycation (Section 4.2).

Insulin resistance is thought to lead to diabetes and is associated with excess body weight gain. (It also occurs temporarily during pregnancy, another time of extra weight gain; see Chapter 6.) The association between weight gain and insulin resistance suggests that expanding adipocytes within growing fat depots alter insulin action, a suggestion explored in the next section.

4.6 Roles of adipose tissue

Adipose tissue (Figure 4.14) consists of adipocytes, connective tissue to hold these cells in place, a blood supply that is shared with adjacent tissues, and innervation from the nervous system (Section 1.2).

Adipose tissue was once thought of as an inert and simply uninteresting long-term storage depot for TAGs.

> Unwanted, unloved, yet often overabundant, few have much regard for fat. Scientists, too, long thought of fat cells as good-for-nothing layabouts unworthy of attention; containers stuffed with energy to be released at the body's command. So stuffed, in fact, that many other parts of the cell were thought too squeezed to function.
>
> Powell (2007)

However, adipose tissue also has a very important role as an endocrine gland, which is explored later in this section.

4.6.1 Adipose tissue as storage

In Chapters 2 and 3, you studied the digestion and absorption of food and the subsequent fate of the macronutrient molecules.

◆ Why is food storage necessary?

◆ You may have suggested a number of reasons, but two important ones are:

1 Food supply in nature is unpredictable. The ability to store excess nutrients and energy allows the individual time to find the next food supply and so confers a survival advantage.

2 It allows the individual time to do other things, like study and sleep!

While the unpredictability of the food supply may have been true for humans living a subsistence lifestyle, it may not be so for life in many modern, developed societies, in which plenty of food of high energy density is readily available.

Humans have evolved ways of storing nutrients in both short- and long-term energy reserves.

◆ What are these two energy stores? Which is long-term and which is short-term?

◆ Glycogen stores form the short-term reserve, and the long-term energy store is in the form of TAGs in adipose tissue (found in various depots throughout the body).

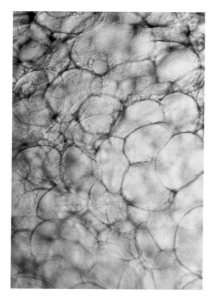

Figure 4.14 Light micrograph (photograph taken through a microscope) of an intact fragment of human adipose tissue.

Table 4.1 lists the various types of energy reserves found in humans and gives an approximate length of time that they can support certain actions. As you can see, TAG stores in adipose tissue are both the largest store and can provide energy over the longest timescale, whatever the level of activity. The amount of TAG quoted in Table 4.1 (Haugen and Drevon, 2007) is for a lean adult, so these values would be larger for obese individuals.

Table 4.1 The various types and typical amounts of energy reserves in humans and how long they can support various activities.

Energy reserve		Time the energy reserve may release sufficient energy for:		
Storage form	Amount/g	inactivity	walking	running
blood and extracellular glucose	20	40 min	15 min	4 min
liver glycogen	80	3.5 h	1 h	18 min
muscle glycogen	350–700	14–28 h	5–10 h	2–3 h
adipose tissue TAG	9000–15 000	34 d	11 d	3 d
protein	6000	15 d	5 d	1.3 d

The volume of individual adipocytes varies markedly, perhaps as much as tenfold, as adult humans become more obese or less obese. The total number of adipocytes found throughout an adult's body is stable in adult life. The primary physical difference between an obese person and one of healthy weight is the *size* of their adipocytes, which is determined by the *quantity* of TAGs stored within them.

So far, we have described the classically accepted storage function of adipocytes. Essentially, these involve:

- absorbing fatty acids that circulate in the blood following a meal
- taking up fatty acids that have been produced from excess glucose taken up by the liver
- the release of fatty acids from adipose tissue as an important source of energy during fasting and intense or protracted exercise.

Adipose depots can be characteristic of the gender (e.g. female breasts, male paunch) and the individual. The typical shape of obese men was previously called android obesity but is now generally referred to as abdominal or central obesity, as shown in Figure 4.15a. Even within the sexes, differences arise: women can have 'apple' or 'pear' body shapes depending on whether there is fat around the abdomen, as in men, or stored subcutaneously in the thighs and buttocks. The latter distribution was originally known as gynoid obesity and is illustrated in Figure 4.15b. The male pattern of obesity with excess abdominal fat is thought to be more associated with adverse risk of comorbidities (Abate and Garg, 1995).

The second role of adipocytes, as producers of hormones and signalling molecules, is the topic of the next section.

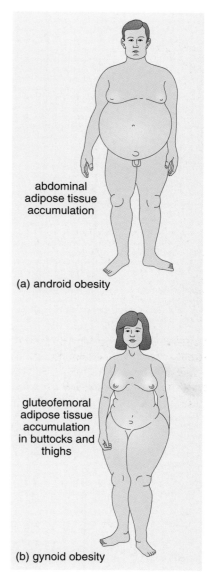

abdominal adipose tissue accumulation

(a) android obesity

gluteofemoral adipose tissue accumulation in buttocks and thighs

(b) gynoid obesity

Figure 4.15 The characteristic shapes of (a) male (android) and (b) female (gynoid) obesity with excess adipose tissue accumulations in the abdominal area and gluteofemoral (buttocks and thighs) areas, respectively.

4.6.2 Stretched to their limits: adipocytes send out distress signals

The 'inert' view of adipose tissue was changed in 1994 with the discovery of the first 'fat messenger' hormone **leptin**, which is secreted from adipocytes. There will be more about leptin in Chapter 5. There have since been further discoveries of signalling molecules secreted by adipose tissue. Adipose tissue is now known to be a dynamic endocrine organ secreting over 50 hormones and other signalling factors, collectively called **adipokines**, that signal changes in adipose tissue mass and regulate fuel usage (Figure 4.16).

These signalling molecules are involved in the regulation of the body's energy reserves by influencing the rates of fatty acid storage and breakdown. We shall briefly consider just three of these signalling molecules here, partly because the precise roles of many of these molecules are not yet known and partly because an overview is sufficient to appreciate the scope and range of their roles.

When adipose cells have capacity for further storage (see Figure 4.17a), they secrete signals such as adiponectin, which improve insulin sensitivity and promote glucose uptake from the blood, leading to its conversion and eventual accumulation as fat. When adipose cells already contain plenty of fat (see Figure 4.17b), they reduce the production of adiponectin. At the same time, they start to produce other signals such as retinol-binding protein 4, RBP4, which acts partly by blocking the action of insulin in muscle (see Figure 4.16 and Yang et al., 2005). That is, RBP4 is one of a number of molecules that initiates insulin resistance, the consequences of which were described earlier. These signals also influence liver and adipose tissue itself. This is a further example of cross-talk between homeostatic systems.

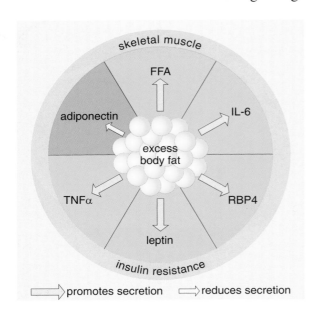

Figure 4.16 Adipose depots that are storing large amounts of body fat secrete hormones and other factors including fatty acids (FFA) that *promote* insulin resistance in other tissues such as skeletal muscle. A few examples are shown. These adipose depots also *reduce* their secretion of adiponectin, which is produced by cells with further storage capacity.

Figure 4.17 (a) Small adipocytes in lean mice send out different chemical signals from (b) large adipocytes in fat mice. Panels (a) and (b) are at the same scale; the adipocytes in (a) are around 100 micrometres in diameter, which means that ten of them (laid side by side) would make a line of cells 1 mm long.

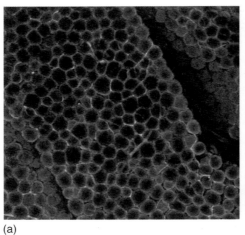

(a)

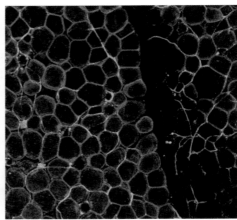

(b)

Adiponectin and RBP4 are produced by and have an effect on adipose depots – as well as other tissues. Other signalling molecules appear to be depot-specific. One of these more specific adipokines is visfatin. Visfatin is preferentially produced by visceral fat as these depots increase in size. It is not produced when subcutaneous fat depots increase in size. Such differentially produced signals may be one way in which different depots are able to change their size at different rates from other depots. The presence of such differentially produced signals also ties in with evidence that *where* fat is found within the body, as well as the rate at which it accumulates, has a bearing on the severity of the health effects caused by excess adipose tissue. For example, not all obese people develop insulin resistance or diabetes (Chapter 7), but those with central obesity appear to be more susceptible than those with excess subcutaneous fat. It is too early to know whether visfatin, or other similar molecules, will have any therapeutic use. The thought is that, if obese people cannot reduce their weight, it might nevertheless be possible to 'interfere' with the signals that their overstretched fat depots are sending out. This is based on the assumption that it is these molecules that are upsetting metabolism that leads to developing comorbidities and that there will not be any undesirable side effects.

A number of other factors start to be secreted by replete (full) adipose cells (for example, resistin, TNF-α and IL-6). As well as promoting insulin resistance, these factors start inflammatory cascades. Inflammation can be thought of as a generalised 'response to damage'. Simple damage, such as a cut finger or contact with stinging nettles, can lead to inflamed tissue, but more extensive inflammation also often precedes disease. Chronic inflammation is thought to underlie several chronic conditions such as arthritis. Thus overstretched adipocytes appear to be behaving as if damaged. One problem very large adipocytes have is getting enough oxygen for respiration. It is thought that this lack of oxygen is partly responsible for the cells behaving as if damaged, triggering the inflammation. The inflammation of adipose tissue could well be contributory in health problems associated with obesity.

4.7   Consolidation

To bring together all the material that we have covered in this chapter, Figure 4.18 summarises the interconnecting pathways of digested macronutrients (carbohydrates, fats and proteins), leading to either their use or storage in energy reserves. This is separated into (1) under conditions of availability (following eating, top panel) and, conversely, (2) the pathways of stored nutrients released from energy reserves for use under conditions of fasting (bottom panel).

Activity 4.2   Energy reserves during feeding and fasting

Allow 10–15 minutes for this activity

Using Figure 4.18, explain where the brain obtains its energy in (a) the fed state and (b) the fasted state.

Now look at the comments on this activity at the end of this book.

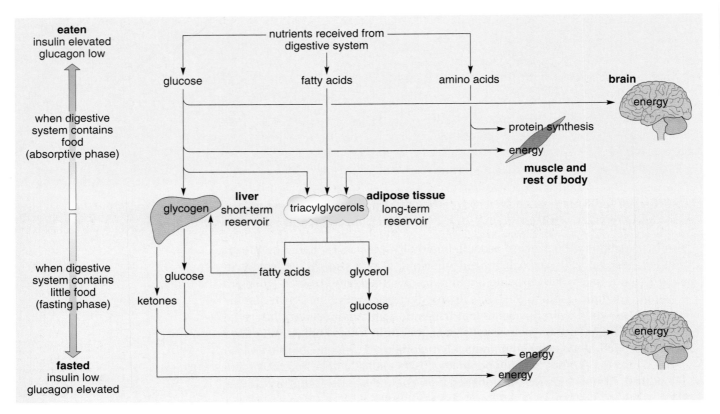

Figure 4.18   Summary of the metabolic pathways involved during the absorptive phase of digestion (glucose, amino acids and fatty acids from the digestive system) when insulin levels are high and glucagon levels are low (upper part of figure) and also during fasting when glucagon levels are high, insulin levels are low and energy reserves release their stores of glucose and fatty acids as metabolic fuels (lower part of figure). During fasting and low levels of insulin secretion, it is mainly the brain and the nervous system that can take up glucose from the blood.

4.8   Summary of Chapter 4

4.1   The body has a continuous need for metabolic fuels, with glucose being a major fuel source for many tissues. Glucose is transported around the body in the bloodstream and its concentration is maintained at optimum levels by homeostasis.

4.2   Insulin is the main hormone regulating the uptake of glucose into the target tissues of liver, skeletal muscle and adipose tissue under well-fed conditions. Glucose may be used or converted and stored for later use. Short-term storage is in the form of glycogen reserves in liver, kidney and muscle; longer-term storage is in the form of TAG in adipose depots throughout the body. Insulin is also involved in the maintenance of adipose depots.

4.3   Glucagon, a hormone with opposing actions to insulin, influences the release from storage of glucose and other metabolic fuels, including fatty acids under fasting conditions.

4.4 Both insulin and glucagon are regulatory factors in glucose homeostasis and maintenance of adipose depots, which is an example of cross-talk between regulatory systems.

4.5 Insulin resistance is a condition that develops when target tissues become less responsive to the actions of insulin. It is linked with increasing body weight and expanding adipose tissue mass, and eventually results in hyperglycaemia (high blood glucose).

4.6 Some tissues – including the brain, nerves, red blood cells, small blood vessels (especially in the eyes) and kidneys – take up glucose independently of insulin. They are susceptible to damage when blood glucose concentrations are high for prolonged periods of time.

4.7 Adipose tissue is important for long-term storage of TAG, with depots throughout the body.

4.8 Adipose tissue is also a dynamic endocrine organ that plays an important role in the body's adaptation to nutritional states. Adipokines, signals that are secreted from adipocytes, are important for normal energy storage in body reserves but are also altered in obesity and implicated in the development of inflammation and disease.

Learning outcomes for Chapter 4

LO 4.1 Define and use, or recognise definitions and applications of, each of the terms printed in **bold** in the text.

LO 4.2 Describe the homeostatic control of glucose levels and the roles of insulin and glucagon.

LO 4.3 Demonstrate an understanding of the metabolic regulation of glucose and fatty acids.

LO 4.4 Outline the role of adipose tissue in storing excess energy in health and disease.

LO 4.5 Describe in simple terms the link between body weight gain and the development of insulin resistance.

Self-assessment questions for Chapter 4

Question 4.1 (LOs 4.1 and 4.2)

What do you understand by the term gluconeogenesis? Consider your answer in terms of the conditions and organs involved.

Question 4.2 (LOs 4.1 and 4.2)

Describe one of the complementary actions of the hormones insulin and glucagon in the regulation of blood glucose concentrations.

Question 4.3 (LO 4.1)

Explain the difference between hypoglycaemia and hyperglycaemia.

Question 4.4 (LO 4.3)

Explain how glucose can become TAG. Under what circumstances does it happen and in which tissue?

Question 4.5 (LOs 4.3, 4.4 and 4.5)

Outline the events that cause an obese individual to develop symptoms such as poor vision.

Question 4.6 (LOs 4.3 and 4.4)

What negative health effects are associated with insulin resistance?

Question 4.7 (LO 4.4)

How do adipocytes signal that they have more storage capacity?

CHAPTER 5

BRAIN AND BEHAVIOUR

En mangeant l'appetit vient. [Appetite comes with eating.]

Old French proverb

5.1 Introduction

In previous chapters, you have seen how increased body weight will inevitably result from a long-term excess of energy intake over energy expenditure. For most (although not all) individuals, this means that they will become obese. Energy intake can be modified by altering eating behaviour. Changes in our energy output are also mainly determined by the choices we make in relation to voluntary exercise. In this chapter, we shall look at the way in which the brain modulates eating behaviour. In particular, we shall consider some of the processes that make us feel hungry before a meal and help us to feel sated after we have eaten. These processes operate at multiple levels. Some involve direct interactions between aspects of our physiology, such as energy availability, and the brain. Others, which are likely to be just as important an influence on eating behaviour and our overall energy balance, depend on psychological and social processes that are heavily dependent on the environment in which we live.

◆ In what situations might increased body weight not be associated with the development of obesity?

◆ Athletes who are exercising heavily will often increase their muscle mass and as a consequence their body weight and BMI (Section 1.3).

Other athletes will deliberately increase their weight as they move from one aspect of their sport to another (Figure 5.1). Although their BMI will increase, they will not satisfy the broader definition of obesity given in Chapter 1.

The relatively extreme example shown in Figure 5.1 confirms a key message for this chapter. Although there are many physiological influences on eating behaviour, it is behavioural mechanisms, extending from relatively simple learning processes through to our sense of who we are and what we want to achieve, that have the major role in whether we are likely to develop increased body weight and obesity.

5.2 The 'eating' brain: a quick introduction

This section provides a brief introduction to brain function, especially to those parts that have a special role in the maintenance of energy balance. The brain, of course, is primarily responsible for the control and production of all human behaviour, including eating. However, it is reliant on signals from elsewhere in the body and from the outside world to help control this process. The net result is an integrated and sophisticated behavioural control system.

Figure 5.1 Ben Ainslie has won three Olympic medals for dinghy sailing. The first (a silver) was in 1996 when he sailed the Laser and weighed 76 kg (height 1.82 m; BMI 22.9). By 2004 he had moved to the Finn class, a larger and heavier boat, and needed to increase his weight to 95 kg (BMI 28.7) in order to be competitive, winning the second of his two gold medals. Note that Ben maintained his athletic build and does not look overweight.

Figure 5.2 shows the major features of the brain. One critical area lies in the base of the brain, or **hindbrain**, not far from the point where it merges into the spinal cord. For an understanding of eating behaviour, note the part of the hindbrain called the **nucleus of the solitary tract**. In addition to being sensitive to hormonal signals relevant to feeding, the nucleus of the solitary tract receives direct neural inputs that play a complementary role to these hormonal signals. For example, signals about stomach distension are transmitted via the **vagus nerve** (which connects organs in the gut, including the stomach and the liver, to the hindbrain) to the nucleus of the solitary tract.

Figure 5.2 View of half the brain, showing some structures relevant to the control of eating and their input from the digestive system.

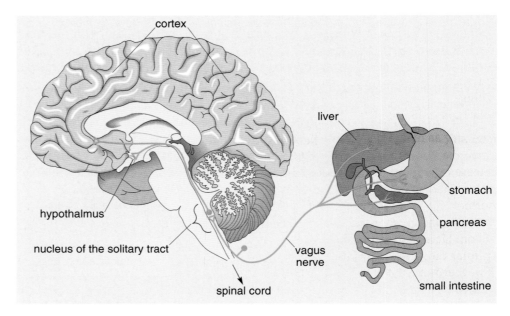

There are neural connections from the nucleus of the solitary tract to a number of other key brain areas for feeding. One of the most important of these is the **hypothalamus** (Figure 5.2). Although the hypothalamus is a relatively small structure, it is a key area – not just for feeding, but also for a number of other processes that are essential to the regulation of human physiology and internal state. These include fluid and salt balance, several aspects of reproductive behaviour and responses to stress. A common feature of all of these functions is that they are critical to both short- and long-term survival. Damage to the hypothalamus can have striking effects on both eating behaviour and, in the longer term, body weight regulation. However, the type of effect depends on which part of the hypothalamus is affected. Overeating and obesity will result when the damage is to areas of the hypothalamus, especially the *ventromedial hypothalamus*, close to the midline of the brain, where the two **cerebral hemispheres** (the left and right halves of the brain) meet. By contrast, damage to the parts of the hypothalamus that are further from the midline, in an area known as the *lateral hypothalamus*, lead to reduced food intake and loss of body weight.

◆ If you were trying to describe the role of the lateral hypothalamus from this observation, would you characterise it as a brain centre that activates or inhibits feeding?

◆ The evidence presented here suggests that the lateral hypothalamus is an area particularly involved in the activation of feeding – a 'feeding centre'.

Although it is sometimes difficult to interpret the effects of brain damage on behaviour, the idea that the lateral hypothalamus has a particular role in generating feeding behaviour is one that was suggested more than 60 years ago and remains influential and well supported to this day.

The hypothalamus has complex sets of connections to other brain areas within the cerebral hemispheres, including structures within the *basal ganglia*, a group of structures that are situated in the middle of the cerebral hemispheres. Although it is a considerable simplification, you can think of the basal ganglia as mostly contributing to the way in which the rewarding properties of food affect eating behaviour. Parts of the **cortex**, which makes up most of the outer, convoluted surface of the cerebral hemispheres, are – at least in the case of eating – more concerned with the complex evaluation of food-related stimuli, including its taste and texture.

The individual neurons that make up the circuits described here communicate with each other at **synapses** through the release of **neurotransmitters**, as shown in Figure 5.3. This process is, in principle, very similar to the way in which hormones act on their target organ. Hormones are released into the bloodstream and carried to the cells that they will act on. Here, hormones bind to receptors and hence affect cellular function. Neurotransmitters work in a similar way. The neurotransmitter is released from the **presynaptic** neuron, diffuses the short distance to the **postsynaptic** neuron and acts on receptors found on the membrane of that cell.

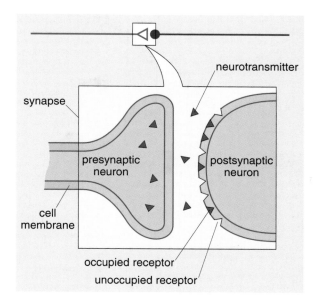

Figure 5.3 Neurotransmitter action at a synapse. The ending of the presynaptic neuron is close to but not in contact with the postsynaptic neuron. In response to a neural signal, molecules of neurotransmitter, shown here as triangles, move out of the presynaptic neuron and bind to their receptors in the cell membrane of the adjacent postsynaptic neuron. This receptor binding initiates a neural signal in the postsynaptic neuron which travels down the neuron away from the synapse.

The next two sections of this chapter will explore the way in which hormones and direct neural inputs relevant to eating reach the brain, and the final section will return to the brain and consider, in greater detail, the contribution of different brain areas to the regulation of eating behaviour and energy balance.

5.3   Hunger

Hunger usually refers to processes that make eating more likely, whereas **satiety** refers to those that make eating likely to stop once it has started. It might seem that satiety is no more than the opposite of hunger – suggesting that eating stops when hunger has diminished. However, because the digestion and absorption of food happens slowly and is not completed until well after we actually stop eating, there are enough differences between hunger and satiety to make it worth considering these two processes separately.

5.3.1   Hunger and an empty stomach

We are all familiar with the often embarrassing sensation of stomach contractions and gurgling – perhaps during an important meeting – before we are due to eat a meal. It is hardly surprising that stomach contractions have sometimes been thought to be the origin of feelings of hunger. Here is an early version of this idea.

We are induced to take food, both from the sense of pain which we call hunger, and from that of the pleasure imparted by the sense of taste. The first of these proceeds undoubtedly from the folds of the stomach, which possess great sensibility, being rubbed against each other, by the peristaltic motion, and by the pressure of the diaphragm and abdominal muscles, so that naked nerves being rubbed against naked nerves excite an intolerable degree of pain. Thus man is both effectually admonished of the dangers of abstinence and excited to procure food by his labours.

von Heller (1765), cited by Teitelbaum (1964)

Although 18th-century physiology does not fit with modern concepts of neuron action, there are several fascinating insights in this quotation from a psychological perspective. First, hunger is seen as an unpleasant state, akin to pain, which evokes appropriate behaviour. Variations of this idea remained popular in psychology until at least the 1960s. However, the quotation also anticipates the idea that eating is jointly determined by physiological need and the rewarding qualities of food – this is a thoroughly modern approach! Present-day psychologists use the term **incentive value** to describe the attractive properties of a particular food, or stimuli associated with it, that lead to it being eaten.

The idea that the stomach was an important source of hunger signals was first investigated experimentally in the early part of the last century by Walter Cannon (Cannon and Washburn, 1912). He persuaded his collaborator, Irving Washburn, to swallow a balloon connected to a tube. Once the balloon was in his stomach, it could be inflated and stomach contractions measured as movements of the air in the tube. Figure 5.4 shows a part of the experimental setup. Cannon claimed that his data supported a correlation between stomach contractions and feelings of hunger. The data show that the contractions usually precede the reporting of hunger. But there are a number of questionable features in this experiment.

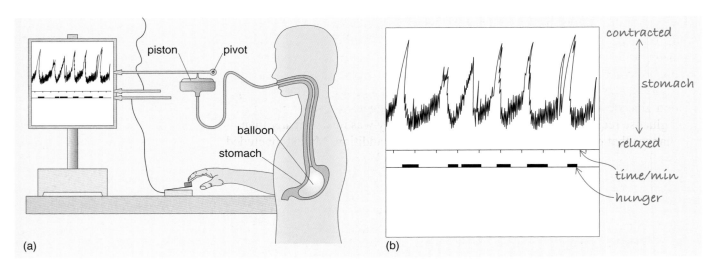

Figure 5.4 Cannon and Washburn's (1912) classic experiment suggesting that stomach contractions, or the associated acid secretion, might provide a stimulus for hunger. (a) The arrangement of the apparatus for the experiment. (b) Data for stomach contractions.

There was only one **participant** (Washburn), and his results might not be representative of other individuals. In general, such experimental work would now only be published if the results had been replicated in a number of individuals.

Washburn was a collaborator of Cannon's and was certainly aware of the hypothesis that they were trying to investigate. This might have influenced, perhaps unconsciously, the way in which he reported his hunger sensations. This problem could be overcome by ensuring that the participants were unaware of the exact purpose of the experiment.

Scientific experiments (and also **clinical trials** of drugs or psychological treatments) in which the participants are unaware of the exact purpose of the experiment or whether they are receiving an active treatment are referred to as *blinded*. If both the participant and the person making the experimental measurements or clinical assessment are unaware of the exact treatment given to a particular participant, then the procedure is described as **double-blind**. Data collected by later investigators did not support the interpretation of Cannon and Washburn's original experiment. However, the stomach and other parts of the gut do play a very important role in relation to both hunger and satiety as a result of the hormones that they secrete into the blood while food is eaten and digested. Some of these were mentioned in Chapter 3.

People who take part in experiments are called participants and today are protected by agreed ethical procedures.

5.3.2 Hunger and blood glucose

Neurons, to a much greater degree than the cells of other tissues such as muscle, are dependent on a steady supply of glucose as their energy source. This, of course, is one reason why glucose levels in the bloodstream are so tightly regulated (Chapter 4). However, blood glucose levels do vary in relation to recent energy intake and it is possible that the changing levels could be used as a signal to influence feeding behaviour. This idea has a long history and perhaps is most often associated with the French–US nutritionist Jean Mayer, who was working and writing in the middle of the 20th century. However, as early as 1916, the Swedish–US physiologist Anton Carlson had suggested that hunger might be associated with a lowering of blood glucose levels. Mayer elaborated the idea into his 'glucostatic' hypothesis which suggested that eating might be regulated by a signal reflecting glucose availability. The simplest signal would be a change in blood glucose level itself. However, the evidence suggested that the change in blood glucose required to induce hunger experimentally was much greater than any changes that occurred under natural conditions. In addition, Mayer concluded that absolute blood glucose levels could not be a controlling variable because people who typically have abnormally high glucose levels, such as those with untreated diabetes, show persistent hunger and weight gain. Instead, he suggested that there might be a mechanism that could compare glucose levels in the blood supply to – and from – the brain. Although this was an ingenious idea, there was no good evidence to support it, and Mayer's views became less influential.

Several decades later, it became possible to measure human blood glucose levels accurately using small, repeated samples in a laboratory situation where the participants would ask for a meal when they felt hungry. The blood samples were taken by inserting an intravenous catheter for the duration of the experiment. There was a clear relationship between a dip in blood glucose level and a request for a meal (Figure 5.5).

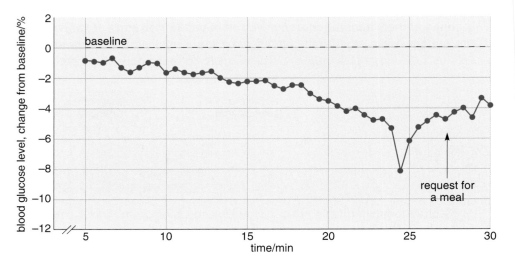

Figure 5.5 Changes in human levels of blood glucose prior to the request for a meal. Note that in this experiment blood samples have been taken every minute, whereas in Figure 4.4 blood samples were only recorded once an hour. In addition, the blood glucose levels are measured as a percentage change from a person's average blood glucose levels. The starting blood glucose level is indicated by the baseline.

◆ Examine Figure 5.5. Describe the relationship between the change in blood glucose and the request for a meal.

◆ There is a transitory dip in blood glucose about 6 minutes before the request for a meal. At the moment that the meal is requested, blood glucose levels have already risen.

The liver is one of the body organs that is sensitive to glucose. It is also able to send neural signals to the brain that reflect the levels of glucose in the hepatic portal vein. This is actually a more *general mechanism* as the liver can also sense the levels of other nutrients, such as fructose from fruit and triacylglycerols (TAGs) from fat, that are being absorbed from the digestive tract and into the blood. In addition, neurons in specific brain areas, including the hypothalamus, may be sensitive to levels of glucose in their blood supply. So glucose levels probably do play a role, as Mayer suggested, but in combination with other factors, such as the gut hormones alluded to earlier.

5.3.3 Gut hormones and hunger

One way in which a physiological state, such as lowered body energy reserves, can affect behaviour is through the release of hormones into the bloodstream. For example, a hormone released from one of the organs of the gut might bind to receptors for that hormone on neurons within the brain. The binding of the hormone to the receptor will lead to a cascade of biochemical changes within the

neuron. In the short term, this may lead to either activation or inhibiti⟨
cells that, in turn, modifies the function of the brain circuits responsibl⟨
(Figure 5.6).

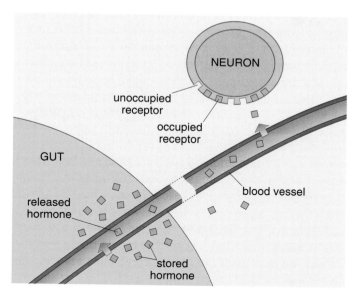

Figure 5.6
of hormone i⟨
the bloodstre⟨
and its effect ⟨
brain. Note tha⟨
principle is sim⟨
to that shown in
Figure 5.3 for th⟨
neurotransmitter
action.

Ghrelin is a hormone discovered in 1999 that may be involved in hunger and
the initiation of eating. It is secreted into the bloodstream by cells in the stomach
and duodenum. Its earliest known role was to modulate the way in which growth
hormone is released from the pituitary gland – hence its strange name, which
is derived from growth-<u>h</u>ormone <u>re</u>leasing factor! However, it was quickly
discovered – first in rodents (rats and mice) and subsequently in humans – that
the administration of ghrelin can act as a powerful stimulant of feeding
behaviour. In addition, repeated measurements of blood ghrelin levels have
demonstrated a close correlation between increases in ghrelin level and eating a
meal. In humans, as Figure 5.7 shows, peaks in ghrelin levels coincide with each
of the main meals of the day.

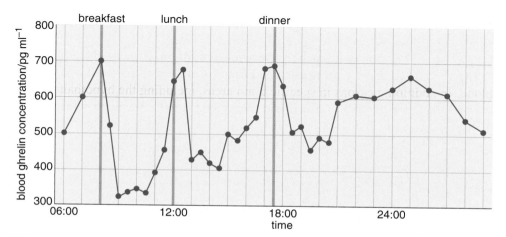

Figure 5.7 Changes in ghrelin associated with human meal taking. The data are
averaged from ten participants who were in an experimental laboratory in which
meals were provided at the times indicated. A picogram (pg) is a millionth of a
millionth of a gram – a very small quantity.

◆ Does the coincidence of eating and ghrelin secretion shown in Figure 5.7 demonstrate that ghrelin is a 'hunger' hormone that determines when you will eat?

◆ No; the correlation of two events with each other can never by itself prove that one causes the other (Section 2.8.2).

◆ Think of an observation from your own experience that might cast some doubt on the hypothesis that ghrelin determines when you will eat.

◆ Imagine a time when you have been strongly engaged in some fascinating task for several hours. You realise that you have missed your regular mealtime yet you do not even feel very hungry. If your ghrelin levels have been increasing since your last meal, then surely this should have led to you breaking off from the task in hand and going off to see what there was to eat.

You might have noticed in Figure 5.7 that there is one early night-time peak in ghrelin levels which is not associated with any meal – presumably because the participants in the experiment were asleep! So there cannot be a direct relationship between ghrelin levels and eating.

There is an additional issue here. Look again at Figure 5.7 and note how rapidly the level of ghrelin increases just before a meal. It is possible that the release of ghrelin close to the time when a meal starts is one part of the preparation that the body makes for the arrival of food and that, at least at the behavioural level, the 'decision' to eat in the near future has already been taken.

Although the discovery of ghrelin and its possible hunger-stimulating effect is a fascinating one, at the time of writing (2008) it is not known how important the hormone is for regulating normal eating. The necessary evidence will probably be available in just a few years because a number of drug companies are developing compounds to block the ghrelin receptor selectively.

◆ What effect would you predict that blocking ghrelin receptors might have on eating (in the short term) and body weight (in the long term)?

◆ The prediction, if there is a causal relationship between increased ghrelin and the initiation of eating, is that blocking ghrelin receptors will reduce feeding behaviour and, with repeated administration, lead to decreased body weight.

A drug that blocks receptors for a given hormone or neurotransmitter is known as an **antagonist**. By contrast, a drug that simply acts like the natural neurotransmitter or hormone is known as an **agonist**. In Chapter 9, we shall take a careful look at the likely utility of ghrelin antagonists and other potential drug treatments for obesity. For the moment, ghrelin remains the only known hormone that might be a natural hunger stimulant. However, there are several other hormones that may act to enhance satiety (see Section 5.4).

5.3.4   Hunger and the liver

In Chapter 2, you were introduced to the role of energy storage of carbohydrate in the form of glycogen within the liver. The liver is also implicated in the control of feeding, although in this case the evidence comes almost exclusively from studies in rodents and it has not yet been possible to collect definitive data in humans. In rats, it is known that the liver can send signals to the brain that reflect energy availability. The evidence comes from studies where the ability of the liver to make ATP from the products of digestion is impaired by the use of compounds that interfere – for a short period of time – with this process. The result of such experiments is that the animals show rapid increases in feeding. This is just what would be expected if one function of the liver is to sense the supply of energy arriving through the hepatic portal vein from the gut. Further studies of this kind have shown that preventing neural signals from passing along the vagus nerve (see Figure 5.2) blocks these experimentally induced changes in feeding behaviour. Given the strong parallels between the physiological mechanisms responsible for energy balance in humans and rodents, it seems very likely that signals from the liver are important in humans as well. Unfortunately, it remains a very difficult area to study experimentally in humans.

5.3.5   Cognitive influences on hunger

Cognition is the general term used by psychologists to describe perception, learning, memory and other processes involving complex mental processing by the brain. Learning may often influence physiological processes, including those involved in the digestion and absorption of a meal, and can help the body to anticipate and prepare for a disturbance in homeostasis that will occur in the near future. This is an example of feedforward (Section 2.8.1).

◆   Why will eating a meal disturb homeostasis?

◆   Because it will result in increases in blood levels of digested nutrients, including glucose. This will activate the feedback mechanisms shown in Figure 4.3.

A well-known preparatory response of this kind is the **preprandial** ('before the meal') release of insulin – clearly shown in Figure 4.4b before dinner is eaten. One obvious explanation for such insulin release is that it prepares the cells of the body to be able to absorb the nutrients that will soon arrive from food. The preprandial release of insulin can be conditioned. This means that, for example, an odour that is associated with eating a food will, in itself, lead to insulin secretion. You may be thinking, quite correctly, that this sounds very reminiscent of the original conditioning experiments carried out by Ivan Pavlov (Box 3.1).

◆   Now look back at Figure 5.5 and, using material from Chapter 4, think of a hypothesis, involving insulin levels, that might explain these data.

◆   A possible hypothesis to explain these data would be that there was a conditioned preprandial pulse of insulin release just before the dip in blood glucose.

There is some evidence for this in rats, but not, so far, in humans.

While some of the cues that lead to hunger and eating may arise from internal physiological signals, there is a great deal of evidence to suggest that environmental cues that have come to be associated with food can enhance appetite. The simplest experiments of this kind were similar to the classic work by Pavlov mentioned previously. Hungry rats received repeated exposures of a stimulus (a buzzer and flashing light) and access to food. A different stimulus was used on an equal number of occasions when food was not available. During the subsequent test phase of the experiment, food was continuously available to the animals. When the buzzer and light were presented the rats rapidly began to feed whereas the alternative stimulus – which provides the **control condition** – had no effect on their behaviour.

Is the same true of human feeding? Do the cues associated with food – the logo used by a fast-food restaurant, the sight of the clock at lunchtime or, for that matter, the sight and smell of a well-cooked meal – arouse our feelings of hunger in an analogous way?

You may feel that these questions are hardly worth asking, because the answers are obviously 'yes'. You would be in good company here. The US psychologist and philosopher William James (brother of Henry James, the novelist) wrote the first classic textbook of psychology in the 1880s. Here is a short extract.

> Not one man in a billion when taking his dinner, ever thinks of utility. He eats because the food tastes good and makes him want more. If you ask him *why* he should eat more of what tastes like that, instead of revering you as a philosopher he will probably laugh at you for a fool.

James (1890)

In recent years, behavioural scientists have collected experimental evidence to support the idea that, in part, we eat because food tastes good rather than because we have a physiological need for it or, in James's rather old-fashioned term, it has utility for us. Figure 5.8 shows one example – the so-called 'appetiser effect'. The participants were invited into the laboratory at lunchtime, having not eaten since breakfast and expecting to receive lunch (and payment – these participants, as in much psychological research, were university students). They were presented with a pasta meal but, before they began to eat, completed ratings of hunger and several other measures. The pasta was presented on a table with a hidden balance. After consumption of every 100 g of food, a participant was asked for additional hunger and fullness ratings. The pasta was provided with a tomato sauce that was differently flavoured for each group of participants. Some received a bland sauce; for others, the flavouring was adjusted to be optimal; and for the final group the sauce flavouring was too strong and judged to be less pleasant. Figure 5.8 shows the changes in hunger ratings over the meal (Yeomans, 1996).

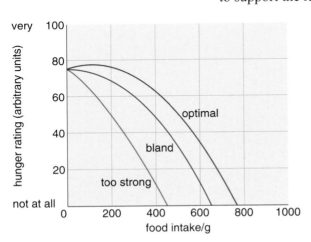

Figure 5.8 Hunger ratings during the course of a meal when the pasta sauce was either bland, optimal or too strong in flavour.

◆ What is the most obvious common feature of these ratings, regardless of the type of sauce provided?

◆ Hunger is high at the beginning of the meal and decreases as the meal progresses.

◆ What is the additional effect on hunger of varying the flavour of the tomato sauce?

◆ When the sauce is optimally flavoured there is a slight additional increase in hunger at the beginning of the meal. When the flavour is either bland or strong enough to be slightly unpleasant, hunger decreases more rapidly.

These pasta meals all tasted relatively good and looked attractive but, as you can see, only an optimal enhancement of flavour increased hunger. Also, and not surprisingly, this condition led to increased total energy intake during the meal. A well-flavoured sauce is not the only way of enhancing appetite and intake. Small amounts of alcohol taken before a meal have also been shown to be highly effective, using a procedure in which the participants were unaware of what they were drinking until the experiment was completed.

5.3.6 Summary

Hunger may be affected by internal physiological signals. It is also strongly influenced by the environment in which we live and the qualities of the food that we are eating. A plentiful supply of tasty, varied food types is very likely to increase food intake in both the short and the long term. Conditioned cues previously associated with eating may arouse our hunger, even when we have no energy deficit. Returning to the message of Chapter 2, long-term enhancement of food intake, if not matched by a similar increase in energy output, will inevitably lead to increased body weight – much of it in the form of fat.

5.4 Satiety

Satiety is the term generally used to describe the state that we 'sink' into after a good meal. Perhaps food is still available but we are no longer motivated to eat, although it may still seem at least potentially attractive. We feel 'full' and perhaps we are also showing other kinds of behaviour (curling up in an armchair, falling asleep). There is nothing uniquely human about this state. The German–US naturalist George Schaller, in his classic book on the behaviour of lions in the Serengeti, describes a very similar sequence after a family group has feasted on a wildebeest (Schaller, 1972), and much the same behaviour has been described in other mammals. So satiety is a state that is easy to recognise at the behavioural level. However, to a greater extent than hunger, it poses real problems at the physiological level. Why is this?

The issue is that eating does not occupy a large fraction of our day – perhaps no more than 5 or 10%. It follows that we take in energy in the form of food much more rapidly than we use it. But digestion of food, and hence the availability of the energy it contains, occurs much more slowly, and mainly after we have already stopped eating. This raises a real problem. If meals serve to correct a current energy deficit, as suggested by homeostatic models (Section 2.8), then how do we know when to stop eating? The experimental evidence suggests that it may be best to think of satiety in terms of a cascade of events which succeed each other and maintain our sated state. We shall use Figure 5.9 as our guide to this cascade.

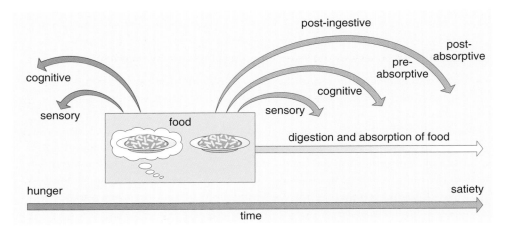

Figure 5.9   The satiety cascade. Some factors enhance hunger (red arrows, pointing to the left): the thought of food and its attractiveness, smell and appearance. However, other factors contribute to satiety (blue arrows, pointing to the right): cognitive factors, i.e. being aware of how much you have eaten; and pre- and post-absorptive factors, from the food and nutrients in the gut and elsewhere.

5.4.1   Cognitive factors in satiety

Cognitive factors, operating in several different ways, may affect the processes that lead us to stop eating towards the end of a meal.

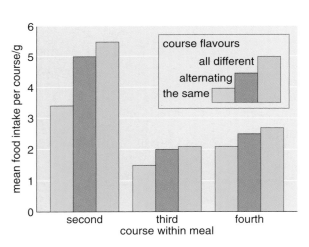

Figure 5.10   Consumption of food when rats were offered four courses of the same flavour (the control condition), two flavours that were alternated or four different flavours. The foods were a semolina-based baby dessert.

Perhaps the simplest cognitive factor relates to the way in which almost all animals show a reduced response to stimuli that are presented on several occasions. This process is known as **habituation**. For example, an unexpected loud noise may make you jump, but the same noise a few seconds later may produce a much smaller response. In the same way, people may rapidly habituate to even the most attractive food if it is repeatedly offered to them. In this context, it is worth noting that one feature of the typical diet in modern urban societies is its variety. There is good evidence to suggest that variety in itself can increase total food intake. As so often, the first formal experiments were done using rats. In one such study, when rats were offered four-course meals that were either of the same flavour or four different flavours, they ate more when there was variety in the meal (Clifton et al., 1987). They showed a similar increase in intake if the meal consisted of two alternating flavours (Figure 5.10). The increase in intake, over the last three courses, was 46% in the full-variety condition and 37% in the alternating condition. These differences (46% and 37%) are not statistically significant. But both values are significantly greater than that obtained from the group of rats who had no variety in their meal. The reduction in intake in the group receiving the unchanging diet is usually described as **sensory-specific satiety**.

In a related experiment (Rolls et al., 1984), which did not include the alternating condition, human participants were presented with four-course meals that either were identical or were of different types of food. In this case, the increased consumption was even greater in the variety condition (60%), perhaps because adult humans would have had a strong cultural expectation that a four-course meal should not consist of the same food presented for each course!

Conditioning experiments have shown that rats and humans, which are both species that eat a wide variety of foods, can learn to associate particular features of a food, such as its taste, with the amount of energy that it contains. Figure 5.11 shows the results of an experiment with human participants in which they arrived for breakfast in the laboratory on a series of days (Yeomans et al., 2005). The experiment investigated whether the participants would change the amount of porridge that they ate depending on its energy content. Two differently flavoured porridges were used. One porridge was high in calories and the other low. They were presented with one type of porridge on the first day and the other on the second day, and ate roughly equal amounts of each. This was just as expected since their palatability had been carefully matched in advance of the experiment. Then, on eight successive training days, they received fixed portions of porridge. On 4 of these days they received the low-calorie porridge and on the other 4 days they received the high-calorie porridge. On two final test days, they could again eat as much porridge as they wished although, as on the first two days, the low-calorie porridge was presented on one day and high-calorie porridge on the other day. Figure 5.11 shows the amount of porridge eaten in terms of its energy content at the beginning and then at the end of the experiment.

◆ What is the effect of training on intake of the less energy-dense porridge?

◆ Consumption of this porridge increases.

Note that the effect is **statistically significant**, indicated by the letters (c and d) above the bars on the right-hand side of the figure. Statistical significance means that a mathematical test has been applied to the experiment's results and has shown there to be a difference that is 'real', i.e. a difference that is most unlikely to have been obtained by chance. When scientists report that the results are *significant*, they mean that they are statistically significant.

◆ Look at Figure 5.11 again. What effect does training have on intake of the more energy-dense porridge?

◆ Consumption of this porridge decreases, as expected. However, in this case the difference is not great enough to be significantly different, indicated by the bars on the left-hand side being labelled with the same letter (a).

An interesting feature of these – and similar – experiments on **conditioned satiety** is that careful questioning and debriefing of the participants at the end of the experiment suggests that they are unaware of what has happened. They may

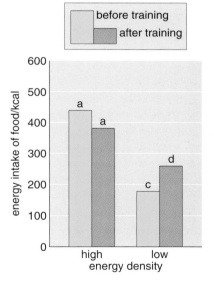

Figure 5.11 Consumption of two distinctively flavoured porridges that differ in energy density before (yellow bar) and after (red bar) training. Groups with the same letter above them are not significantly different from one another.

have learnt the flavour–calorie relationship, but they don't know it! Psychologists refer to this kind of learning without awareness as **implicit learning**. It is likely that such processes play a major part in learning about the properties of foods and their consequences. As you will see at the end of Chapter 8, it may be one of the reasons why attempts to diet as a strategy for reducing body weight are relatively unsuccessful.

5.4.2 Pre-absorptive factors and satiety: a full stomach?

There is some evidence to suggest that signals from the stomach, related to both the energy content and the volume of food consumed, may influence the development of satiety. Signals of this kind are referred to as *pre-absorptive* (see Figure 5.9) because they are generated before food has been absorbed from the digestive tract into the bloodstream. In one experiment, Barbara Rolls and her colleagues used a tube, which ran into the body through the nose, down the oesophagus and into the stomach, to *infuse* liquid into the stomachs of their participants (Kral and Rolls, 2004). The liquid varied in both volume and energy content. On one occasion, although it appeared to the participants the liquid was infused, it was actually diverted without them being aware of this. After 30 minutes, the participants had free access to lunch.

◆ What effect would you expect the infusions to have?

◆ The infusions should reduce intake, and the effect should be greater with increasing volume or increasing energy content of the infusion.

In fact, this is exactly the result that was observed.

◆ Why was the last experimental condition (tube inserted but no liquid infused) included?

◆ It provided a *control* condition, which helps to assess whether there are any effects of simply inserting the tube on the amount of food that is subsequently consumed during lunch.

◆ Which body organs might be providing the signals that led to these reductions in intake?

◆ There are a number of possibilities. The signals might originate from the stomach itself, but it is also possible that they are produced when the infused liquid moves on into the duodenum or, in the case of energy-related signals, after some absorption has taken place.

It has not been possible to separate these possibilities in humans. Experiments in rats, in which, in addition to the possibility of infusing fluid into the stomach, the exit from the stomach could be temporarily closed, have suggested that direct signals from the stomach may be of importance. However, a substantial part of the effect is likely to result from signals generated after the food has left the stomach.

5.4.3   Pre-absorptive factors: hormones and satiety

A number of hormones are released shortly after eating. These hormones might also act as useful indicators of energy intake, effectively acting as additional pre-absorptive satiety signals. There is excellent evidence in favour of at least some hormones acting in this way. In Chapters 3 and 4, you learnt that insulin is released as part of the physiological response to increasing blood glucose levels. The natural release of insulin in the context of a meal leads to eating less and, as you shall see later in this chapter, this action seems to be mediated by the brain. Cholecystokinin (CCK) is another hormone that is implicated in the process of satiation (Section 3.3).

◆   When and from where is CCK released into the blood?

◆   CCK is released into the blood when the stomach contents move into the duodenum. It is released from cells in the wall of the duodenum in response to the presence of fatty acids.

CCK stimulates the release of bile from the gall bladder into the gut which serves to neutralise the stomach acid so that the next stages of digestion can take place. But CCK may also act as a satiety factor. There are multiple lines of evidence to support this idea. The increased CCK level in the blood is closely correlated with the onset of feelings of satiety. But, of course, a correlation of this kind cannot show that one event (hormone release) causes another (satiety). However, it is also known that the administration of synthetic CCK reduces the amount eaten and that this effect is apparently not caused by any secondary consequences, such as feeling sick. An even more convincing piece of evidence is provided by the use of drugs that block the effects of CCK at its receptors (CCK antagonists) or by studying the effect of genetic mutations of the gene that codes for the CCK receptor. The drugs – at least in experimental studies in animals – can enhance food intake in the short term and lead to raised body weight in the longer term.

◆   What should be the consequences of a mutation that results in CCK no longer activating its receptor?

◆   The effects should be similar to long-term treatment with a CCK antagonist. Food intake and body weight should increase.

Exactly this result has been observed in experimental studies using rodents.

Given these converging pieces of evidence, you might think that CCK – or a drug that mimics its effects – would be a prime candidate to enhance satiety and reduce food intake, and therefore to have use in the treatment of obesity. This idea will be examined in Chapter 9.

CCK is not the only hormone with a potential role in the development of satiety after a meal. Evidence has been published suggesting the involvement of at least three other hormones.

1   Glucagon-like peptide 1 (GLP-1) is a hormone that is released into the bloodstream by cells in the gut in response to the presence of food. GLP-1 can stimulate insulin release from the pancreas. In humans, the infusion

of GLP-1 can reduce food intake without apparently causing feelings of sickness. However, the doses that need to be given produce rather higher levels of GLP-1 than are normally found after a meal.

◈ How (theoretically at least) would you demonstrate that GLP-1 produced within the human body plays a role in satiety?

◆ By administering a GLP-1 antagonist and observing an increase in food intake during a subsequent test meal.

Unfortunately, a chemical of this kind is not available at present.

2   Peptide YY (PYY) is a hormone which is released into the blood from cells in the small intestine in response to the presence of food. Infusion of this hormone into the bloodstream also reduces eating in humans.

3   **Amylin** is a hormone released from the pancreas. Amylin levels may be low in people with diabetes and there is evidence to suggest that treatment with a drug called *pramlintide*, which mimics the effects of amylin, reduces the weight gain that is often evident in this condition.

5.4.4   Post-absorptive factors and satiety: the liver

We have now considered at least some of the short term, pre-absorptive responses to food intake that may lead to satiety (Figure 5.9). Figure 5.9 also suggests that there may be post-absorptive signals that are directly related to the delivery of nutrients from the gut to the rest of the body, and that these act as the final components of the satiety cascade. In Chapter 3, you looked at the way in which complex food molecules are broken down into the basic building blocks of metabolism. These are monosaccharides (such as glucose), amino acids from proteins and fatty acids and glycerol from fats. All of these molecules can enter the basic energy-producing metabolic pathways that you met in Chapter 2. A fraction will be directly used to produce chemical energy in the form of ATP, whereas the remainder may enter short- or longer-term storage in the liver, muscles or adipose tissue. The liver is in a particularly useful position to be able to sense these incoming energy supplies. Nutrients are absorbed from the digestive tract into the blood capillaries that drain into the hepatic portal vein which runs directly to the liver. Experimental studies show that glucose from a meal high in carbohydrate can begin to arrive in the liver within 5 to 7 minutes of that meal starting, although digestion will not be complete until much later. Within liver cells, there are enzymes – one of which is known as **AMP kinase** – that are highly sensitive to the availability of ATP. The resulting changes in the properties of this enzyme may then affect the activity of neurons in the vagus nerve (Figure 5.2), thereby signalling energy availability in the liver.

5.4.5   Adipose tissue and leptin

Hormonal signals are also significant in the later, post-absorptive phase of a meal. The best known of these hormones is leptin. Leptin is produced in adipose tissue and the levels of this hormone in the bloodstream are strongly correlated with a person's quantity of adipose tissue. Thus the level of leptin in the blood provides a signal of the amount of adipose tissue in the body. Animals, human or rodent, that are unable to produce either leptin or its receptors – because of a genetic mutation – show considerable increases in food intake and develop very severe obesity. Humans who cannot produce leptin also show disturbances in growth patterns and reproductive function. All of these problems can be overcome by treatment with a synthetic version of the hormone. However, the number of individuals worldwide who cannot make active leptin or its receptors is very small. There will be more in Chapter 6 about genetic influences on the development of obesity.

◆ What will be the effect of leptin treatment in a person who cannot make receptors for leptin?

◆ It will have no effect. If a hormone signalling pathway is to be functional then the receptor as well as the chemical signal must be present.

Most of the evidence suggests that leptin plays only a small role in meal-to-meal regulation of eating – its importance lies in its role as a longer-term modulator of food intake. In addition, obese individuals with high levels of circulating leptin become insensitive to its effects. This process is similar to the development of insensitivity to the effects of insulin discussed in Chapter 4.

5.4.6 Do foods vary in their satiating capacity?

It has often been suggested that foods vary in their satiating capacity. In particular, some evidence suggests that fats may be only weakly satiating, despite their ability to slow the rate at which the contents of the stomach pass into the duodenum. It has also been suggested that energy-providing drinks, especially those containing sugars or alcohol, may be weakly satiating. Of course, at first sight, this seems inconsistent with the idea that energy availability acts as an ultimate – but critical – satiety signal. However, as you saw in Chapter 1, the last few thousand years and, more especially, the last 50 or so years in modern urban societies has been accompanied by a shift towards a diet of higher energy density, a higher proportion of fat and a marked increase in the consumption of sweet, energy-containing drinks. These usually contain high levels of carbohydrate: sometimes glucose or sucrose, but also often a syrup derived from corn (modified corn syrup), which contains high levels of fructose. Alcoholic drinks may also make a significant contribution to some individuals' energy budgets. It is not surprising that it has been suggested that these factors are an important component of our present-day obesogenic environment.

However, this kind of correlation does not, of course, indicate that one or other of these dietary changes has been critical in increasing the incidence of obesity. One way of trying to resolve this issue in the laboratory is to use a method known as the *preload technique*. Participants are invited to an eating laboratory and are first asked to consume a drink. The drinks that are provided contain energy in different forms and amounts. This is the preload. A few minutes later, participants are provided with free access to food and their intake is carefully measured. The question of interest is the extent to which changing the energy content or macronutrient composition of the preload will lead to a compensatory change in energy intake when food is provided.

◆ Participants in an experiment are given one of two preloads which differed in energy content by 100 kcal. A little later, they are allowed free access to a meal. What difference in energy intake during this meal might you expect?

◆ If the participants compensate for the difference in energy content in the preload, then those given the high-energy preload should eat 100 kcal less in the subsequent meal.

Perhaps unexpectedly, the participants in such experiments often *fail* to compensate fully for the energy that they consumed in the preload by appropriately reducing their food intake. The type of nutrient in the preload has little effect on the extent to which the participants fail to compensate.

Other studies have looked at the role of dietary fat. In one experiment, the participants were individuals who habitually ate a high-fat diet (Blundell et al., 2005). Among this group, some had become obese, whereas others had remained lean. When they were each provided with a test meal, it was clear that the obese individuals had a weaker satiety response to fat. In other words, they reported that the meal had not satisfied their hunger. One possibility, which has yet to be tested, is that these individuals might also show a weaker hormonal response to the meal.

Longer-term studies are also consistent with the idea that humans are poor at detecting disguised changes in the energy density of their diet and then compensating by either reducing or increasing the amount that is eaten. In one experiment of this type, the participants were divided into three groups. For 14 days, each group received a diet of different fat content (Poppitt and Prentice, 1996). The other constituents of the diet, which were mainly protein and carbohydrate, were varied in such a way as to keep the overall energy density of the diets constant. The participants in this experiment were not aware that their diets were being manipulated in this way. In practical terms, this kind of manipulation is achieved by substituting fat for carbohydrate within a yogurt while, at the same time, ensuring that the sensory characteristics (e.g. taste, smell and feel) do not change. The results of this experiment are shown in Figure 5.12a.

Figure 5.12 Energy intake as a function of dietary energy density in two separate experiments, each involving three groups of participants. In (a), fat content was increased and, at the same time, the other constituents were manipulated so that the overall energy density of the diet remained constant. In (b), fat was added, but no attempt was made to control energy density by changing the other constituents of the diet.

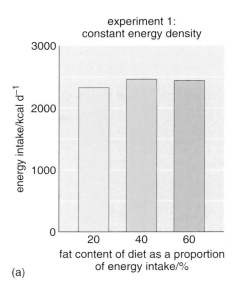

(a)

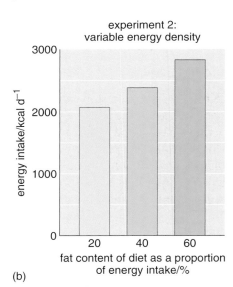

(b)

◆ Look at Figure 5.12a. How did the voluntary energy intake of the participants vary as a function of the fat content of the diets when energy density is constant?

◆ The overall result was straightforward: the energy intake of the volunteers hardly varied as a function of the amount of fat in the diet.

The apparent difference between the 20% and 40% groups in Figure 5.12a was not statistically significant. In addition, there was no significant change in body weight over a period of a few days.

In a second experiment, the participants were again divided into three groups. Now each group received a diet in which the fat content had been covertly manipulated so as to vary the total energy density. One group received a diet which was relatively low-fat and low-energy; for a second group, it was relatively normal; and the third group received an energy-dense, high-fat diet. Other factors, such as the palatability of the foods provided, were again held constant. Surprisingly, the participants in the different experimental groups ate about the same quantity of food (in grams) each day *regardless* of its energy content. The results of this experiment – again in terms of energy intake – are shown in Figure 5.12b.

◆ Look at Figure 5.12b. What was the consequence of this behaviour for the energy intake of the participants?

◆ The consequence was that participants in the high energy density and fat content group had a higher energy intake than those in the groups provided with foods of lower energy density and fat content.

One result of these different energy intakes was that participants in the high-fat group increased in weight during the study period, whereas those in the low-fat group lost weight.

The important message from these two experiments is that humans do not respond very accurately to changes in the energy density of their diet, especially when it is high in fat.

5.4.7 Summary

'Knowing' when to stop eating poses a considerable problem at the level of both physiology and behaviour. The problem arises because the energy that is available in food only becomes completely available well after eating has stopped. It seems that the problem is solved by using a series of different signals that help to predict the extent to which energy reserves have been replenished by the food that has been eaten. None of these signals is sufficient by itself, but together they provide reasonable meal-to-meal regulation of energy balance. However, as you have seen from some of the experimental evidence, there are some high-energy foods and drinks that may provide less effective cues than others. These are the very foods whose consumption has greatly increased in the last few decades and are likely to be important components of our present-day obesogenic environment.

5.5   Brain mechanisms and feeding behaviour

The two previous sections of this chapter have described the importance of signals that relate to both internal state and external stimuli in modifying our eating behaviour. In this section, we are going to look in more detail at some of the key brain structures that are involved in the processing of these signals.

5.5.1   Hindbrain and hypothalamus

Several structures in the hindbrain – most importantly, the nucleus of the solitary tract – receive both hormonal and neural inputs that are relevant to current energy balance. Receptors for hormones such as CCK and leptin are present in this and surrounding areas. In addition, as briefly mentioned in Section 5.2, there are neural inputs, primarily from the vagus nerve, that originate from the stomach and liver that may signal factors such as stomach distension and the rate of energy supply to the liver. These different hormonal and neural signals may interact within the nucleus of the solitary tract and the resultant outputs affect the activity of other brain areas, including the hypothalamus (Figure 5.2).

The hypothalamus is known to be an especially important area for the regulation of eating (Section 5.2). It can be divided into a number of subareas. These subareas are referred to as nuclei (e.g. the **arcuate nucleus**) and the nuclei are grouped into areas on the basis of their position. Thus the lateral hypothalamus is towards the side of the hypothalamus and the ventromedial hypothalamus is towards the middle ('medial') and bottom ('ventral') of the hypothalamus. There are a number of hypothalamic areas that are critical to the control of feeding behaviour and energy balance. The arcuate nucleus is situated at the very bottom of the ventromedial hypothalamus and is also very close to one of the fluid-filled areas, called ventricles, within the brain. The ventricles are lined with blood vessels, so blood-borne signals, such as hormones, can easily diffuse into the cerebrospinal fluid in the ventricles. The arcuate nucleus contains a number of different cells that can be distinguished on the basis of the types of neurotransmitters that they produce. One cell type produces a neurotransmitter known as **neuropeptide Y** (**NPY**; not be confused with peptide YY (Section 5.4.3), despite the similar name). The activity level of NPY-containing cells is determined by a number of different types of receptors that are found on their cell membranes. These include receptors for leptin and insulin, as well as the neurotransmitter **serotonin** (Figure 5.13). These NPY-containing cells influence a number of other areas in the brain, including the lateral hypothalamus. This area has often been referred to as the 'feeding centre' of the hypothalamus (Section 5.2) and has extensive connections with brain areas that organise the physical activities that are required for the ingestion of food. Activation of the NPY-containing cells in the arcuate nucleus leads to a stimulation of 'eating neurons' and thereby stimulates eating behaviour.

There is a second group of cells within the arcuate nucleus which releases a different peptide neurotransmitter known as α**MSH**. These cells, shown in Figure 5.13, influence similar brain areas to the NPY-containing cells but have the *opposite* function in relation to feeding behaviour – they inhibit it. As with the NPY-containing cells, these αMSH-containing cells are sensitive to a variety of chemical signals that are relevant to nutrient availability, including leptin and

In brain anatomy, the term *nucleus* describes a group of neurons. Be careful not to confuse this with the nucleus of a single cell.

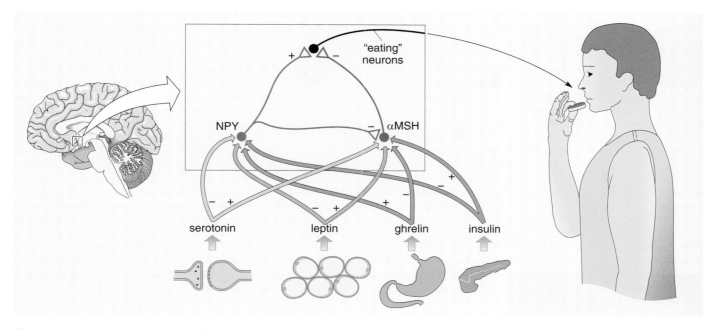

Figure 5.13 Sources of signalling molecules and their influences on NPY- and αMSH-containing cells within the arcuate nucleus of the hypothalamus.

insulin. Leptin and insulin levels will both increase as food is absorbed after a meal. Indeed, in the case of insulin, its release may even precede the meal as a result of a learnt anticipation (Section 5.3.5).

◆ αMSH-containing cells inhibit hypothalamic output cells ('eating neurons') that stimulate eating. What effect should leptin and insulin have on the activity of the αMSH-containing cells?

◆ The αMSH-containing cells should be activated by leptin and insulin.

This is exactly what has been observed and is shown in Figure 5.13. Serotonin, a neurotransmitter which is associated with the short-term development of satiety, has a similar effect to leptin on these αMSH-containing cells. Ghrelin, by contrast, inhibits the activity of the αMSH-containing cells. As Figure 5.13 illustrates, the effects of leptin, insulin, ghrelin and serotonin on NPY-containing cells is the opposite of the effects on αMSH-containing cells.

There is one additional, and very interesting, feature of the model illustrated in Figure 5.13. NPY-containing cells are able to inhibit αMSH-containing cells, but the reverse is not true. It has been argued that this biases the neural control system for eating towards excess consumption. It is easy to see that this might be advantageous in an environment in which food is short, but much less useful in a world in which food is abundant.

5.5.2 Cortical areas and feeding

Thus it seems that the hypothalamus and some areas of the hindbrain act as the critical inputs for information related to current energy reserves, both from hormones released into the bloodstream, often as a part of the process of dealing with food during digestion, and also directly from nervous system inputs from structures such as the liver and gut. But if the hypothalamus and its connections with the brain stem are responsible for initiating responses to the state of our energy reserves, then how do the powerful effects of incentive and the attractiveness of food interact with this system? Imaging studies using brain scanners to measure the activity of particular brain regions have not been very helpful in relation to the hypothalamus because its small size makes it difficult to separate from surrounding areas. However, such imaging studies have been very much more useful in relation to the areas of the brain involved with the rewarding properties of food.

The following study (McCabe and Rolls, 2007) not only demonstrates the way in which taste and smell activate areas of the human cortex but also shows how these factors may interact in rather unexpected ways. In this experiment, the participants lay in a brain scanner and were exposed to various odours and tastes. Included in these was the taste of monosodium glutamate (MSG), originally recognised as an important flavour component of Chinese and Japanese cuisines. It is now recognised as the fifth fundamental taste in addition to salt, sour, sweet and bitter and is usually termed **umami** and, by itself, is sometimes described as 'meaty' or 'savoury'. MSG is also found in high concentrations in Parmesan and Roquefort cheese, and is now extremely widely used in the 'fast food' industry. Returning to the experiment, the participants were also exposed to a vegetable odour as well as the taste of MSG. When the stimuli were presented separately, umami was rated of neutral pleasantness, whereas the vegetable odour was rated as rather unpleasant. However, the combination of MSG and vegetable odour was rated as very pleasant. In addition, as shown in Figure 5.14, the combination also produced a striking activation of the orbitofrontal cortex. Both MSG and the vegetable odour were very much less effective when presented separately.

◆ What do these data suggest about the function of this area of the brain?

◆ It must be an area where taste and odour information come together and are evaluated for their attractiveness.

5.5.3   Brain reward pathways and feeding

Very palatable foods, especially those high in fat and carbohydrate, may be potent stimuli for neural pathways that also respond to different types of rewarding stimuli. Direct evidence for this idea has been found in studies of non-human animals. For example, it is known that many types of both natural (e.g. food, sex) and drug (e.g. cocaine, amphetamine, opiates, nicotine) rewards are able to stimulate activity within a brain structure known as the ventral striatum. The neurons in this area respond to the neurotransmitter **dopamine**. When a rat eats a highly palatable and very sweet breakfast cereal, there is an almost immediate release of dopamine within the ventral striatum. Other foods high in fat and carbohydrate are known to produce the same effect.

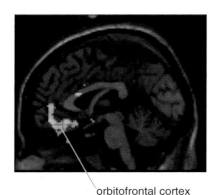
orbitofrontal cortex

Figure 5.14   Activation of the orbitofrontal cortex (yellow/orange, towards the left-hand side) by the combination of MSG (taste) and vegetable (odour).

There is preliminary evidence that the same mechanism may be involved in human responses to food-related cues. In 2001, Nora Volkow and her colleagues used brain imaging techniques to assess dopamine receptor availability in the brains of both healthy weight and morbidly obese individuals (Figure 5.15; Wang et al., 2001). They found significantly lowered dopamine receptor availability in the ventral striatum of the obese individuals.

◆ Use Figure 5.15 to describe the relationship between the degree of obesity and dopamine receptor availability.

◆ Low dopamine receptor availability was associated with a high level of obesity.

Where one variable decreases (in this case, dopamine receptor availability) as another variable increases (in this case, BMI), there is said to be a **negative correlation**. (Compare this with positive correlation in Section 2.8.2.)

◆ If dopamine receptor availability is lowered in obesity, what might be the effect on the functioning of the ventral striatum?

◆ The cells in the ventral striatum will produce a smaller response to food than normal.

Volkow and her colleagues reasoned that the lowered dopamine receptor availability might generate pathological increases in eating as a compensatory response. Interestingly, similar decreases in dopamine receptor availability have been observed in the brains of people addicted to cocaine. However, it is not possible to say whether the lowered dopamine receptor availability was the original cause of a pathological increase in eating or drug use, or simply an adaptive response that only occurred as a consequence of the individual's behaviour.

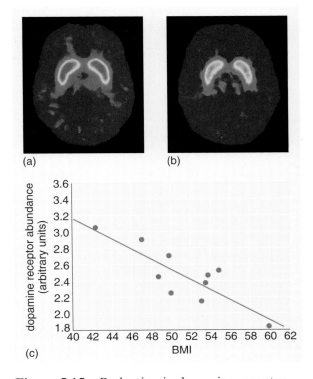

(a)   (b)

(c)

Figure 5.15 Reduction in dopamine receptor abundance in obese individuals. The top panel shows the dopamine receptor availability in the ventral striatum (orange/red areas towards the centre of the brain scans) of (a) control and (b) obese individuals. (c) The negative correlation between dopamine receptor availability and degree of obesity in the ten obese individuals.

5.5.4 Summary

In this section we have looked at some of the brain mechanisms that are involved in our responses to food. The first section of the chapter suggested that eating is jointly determined by internal physiological signals that relate to the availability of energy in the body and the incentive value of available foods and stimuli associated with them. In simpler terms, we eat both because we *need* food and because it is *attractive* to us. Reflecting this dichotomy, there are some areas of the brain, such as the hypothalamus, which may be more strongly responding to physiological need and others, including the dopamine system and its cortical connections, which are more strongly associated with incentive value and reward. However, the brain circuits for eating are highly integrated, so these distinctions should not be overemphasised.

5.6 Summary of Chapter 5

5.1 The control of eating can be thought of in terms of factors affecting the complementary processes of hunger and satiety.

5.2 Factors that may influence hunger include energy availability (specifically, the availability of glucose), the hormone ghrelin and a variety of cognitive influences.

5.3 The most important cognitive influences on hunger are likely to be stimuli that we associate with food, the variety of foods available to us and the extensive use of flavour enhancers. These different stimuli are one aspect of the modern obesogenic environment.

5.4 Satiety is also affected by a variety of different hormonal and neural signals that become important as we eat and digest a meal.

5.5 A number of brain areas form an integrated network to control eating.

5.6 Important brain areas for eating include the nucleus of the solitary tract, the hypothalamus, the orbitofrontal cortex and the dopamine system.

Learning outcomes for Chapter 5

LO 5.1 Define and use, or recognise definitions and applications of, each of the terms printed in **bold** in the text.

LO 5.2 Outline the different physiological and behavioural influences on hunger.

LO 5.3 Outline the different physiological and behavioural influences on satiety.

LO 5.4 Describe, in simple terms, the contribution of different brain areas to the processes that modulate hunger and satiety.

Self-assessment questions for Chapter 5

Question 5.1 (LOs 5.1, 5.2 and 5.3)

How are the concepts of hunger and satiety helpful in understanding the control of eating behaviour?

Question 5.2 (LOs 5.1 and 5.2)

A new hormone is discovered and, as with ghrelin, increases in blood levels are found to correlate with periods of eating. Describe the further experiments that would help to establish whether this hormone is directly involved in stimulating eating.

Question 5.3 (LOs 5.1, 5.2, 5.3 and 5.4)

Briefly explain how neurons within the arcuate nucleus of the hypothalamus integrate the different signals that they receive so that they can help to control eating.

Question 5.4 (LOs 5.1, 5.2, 5.3 and 5.4)

How has the study of brain and behaviour in relation to eating extended our knowledge in relation to the modern-day obesogenic environment?

CHAPTER 6

INDIVIDUAL DIFFERENCES: GENES AND ENVIRONMENT

> In some families the individuals fatten readily; in others with difficulty.
>
> Davenport (1923)

6.1 Introduction

In previous chapters, you have read about how we obtain, digest and use food, and might then store excess energy as fat. But a critical question in relation to obesity is why people differ so greatly with regards to their individual body shape and size (Figure 6.1).

Figure 6.1 People come in many different body shapes and sizes.

◆ What factors are apparent from Figure 6.1 that might affect shape and size?

◆ You might have noticed that the people vary according to age, gender and ethnicity.

Later in this chapter we shall examine these possible contributors to individual differences in shape and size in more detail. To begin with, however, it is necessary to consider two variables that, interacting with each other, have the potential to account for the huge range of individual diversity in body mass:

(1) genetic information inherited from our parents ('genes' for short)

(2) environmental factors.

Later in this chapter you will see how the genetic contribution to body size can be recognised by observing the striking similarities of identical twins. Environmental factors, in this context, means *all* the physical and social variables that interact with our genes during our lifetime, including the uterus (womb) in which the growing baby develops, the number of children in the family into which the child is born, the opportunities for physical activity and the diet that is available.

Genes and environment interact in complex ways. A dramatic example comes from studies of native Pima Indians, either living their traditional lifestyle as ranchers and subsistence farmers in the remote Sierra Madre mountains of Mexico, or residing in more urban environments in the USA (Schulz et al., 2006). In Mexico, their rates of obesity and diabetes are low and are no different from other residents of the Sierra Madre. However, individuals who are genetically very similar, but have migrated to the cities of Arizona in the USA, have obesity rates that are significantly higher than their compatriots in Mexico (10 times higher in men, and 3 times higher in women) and are also higher than those of other residents of Arizona. The prevalence of diabetes in this group is among the highest ever recorded. It is clear that Pima Indians have a genetic makeup that makes them highly susceptible to the effects of the obesogenic US environment. It is equally clear that this genetic susceptibility does *not* mean that they will inevitably become obese. In the Sierra Madre, their lifestyle, which includes high levels of physical activity and a lower-fat diet, has a protective effect against the development of both obesity and diabetes. Change the environment, however, and you may change the response, as this example shows. In broad terms, **genetic susceptibility** means that an individual's genetic makeup makes a given response more likely under some (but not all) environmental conditions.

In this chapter, you will focus on individual differences to see the different ways in which genes and environment influence obesity. Then you will explore the early stages of the human life cycle, and some of the factors that increase the chances of becoming overweight and obese.

6.2 Genes and obesity

There are important genetic influences on being either lean or overweight, and in rare cases, on developing overt obesity. We will look at the role genes can play in influencing individual shape and size after considering some basic genetic concepts that are relevant to weight gain in individuals.

◆ What function (job) do genes perform?

◆ Genes provide the instructions for the sequence in which amino acids are to be linked together to form proteins (Section 2.4.3).

Each gene provides the instructions for a particular protein to be made within a cell that is expressing that gene. Some people have slightly different variants of a protein because **gene variants** have arisen from time to time through mutations. Although many gene variants have no apparent effect, some may modify the function of the resulting protein or, more rarely, may prevent it from functioning at all. A different change to the same gene results in another distinct gene variant. Mutations occur infrequently and only spread slowly through a population, so it has taken many, many generations for the number of variants of each gene present in the population today to arise. The actual gene variants we inherit from our parents is a matter of chance, depending in part on what gene variants they inherited from their own parents.

It is only when a gene becomes active, sending messenger molecules into the cytosol where they are used to manufacture the relevant protein, that we say that the gene is being expressed (Section 2.4.3).

Spontaneous mutations in a particular gene may occur rarely. If the mutation occurs in the cells that generate sperms or eggs, this will result in a child with a gene variant not found in either parent.

We inherit one full set of genes, contained in 23 chromosomes, from each parent. Most cells of our body then contain 46 chromosomes and hence two copies of each gene. One consequence of there being gene variants and of inheriting genes in pairs is that for some genes we inherit the same gene variants from each parent, whereas for other genes we inherit different gene variants from each parent. And for each of us, with the exception of *monozygotic* (identical) twins, the combination of gene variants we inherit is different from the combination of gene variants inherited by anyone else. Our different combinations of gene variants, through their effect on the resulting proteins, is one source of the differences between us.

A second consequence of there being gene variants and of inheriting genes in pairs is that a particular feature, such as morbid obesity caused by untreated leptin deficiency (as described in Section 5.4.5), may appear in a child by inheritance although neither parent was morbidly obese. And these same parents could also have other children who were not morbidly obese. The explanation for such cases is that each parent has one faulty gene variant of the leptin gene (i.e. a gene variant whose instructions result in a faulty leptin molecule) and one normal gene variant of the leptin gene. Some of their children, by chance, receive from each parent the chromosome containing the faulty gene variant. These children with two copies of the faulty gene variant have leptin deficiency and develop obesity. Other children may have the faulty gene variant on only one chromosome (and a normal gene variant on the other chromosome). This pairing is the same as that of the parents, and leads to decreased levels of leptin in the bloodstream and an increased probability of obesity, although to a very much lesser extent than when an individual inherits two copies of the faulty gene variant. A third group of children may inherit two normal leptin genes and have a lower probability of becoming obese than their parents.

◆ Identical twins have an identical combination of gene variants because they arise from the splitting of a single fertilised egg (zygote) and are the exception to the rule that each person has a unique combination of gene variants.

◆ Will these children with two normal leptin genes pass on the possibility of becoming obese to their offspring?

◆ No; each individual has only two copies of the gene for leptin, so these children have no leptin deficiency variants of the leptin gene and they will not pass on the leptin deficiency to their offspring. If their offspring become obese, it will be for other reasons.

Faulty gene variants of the leptin gene (or the leptin receptor gene) result in early onset morbid obesity, but are very rare. These cases may be referred to as **monogenic** obesity, because a change in a *single* gene, by itself, has such a noticeable effect. However, very few cases of obesity can be explained as the result of a single faulty gene.

◆ Why is monogenic obesity rare?

◆ It is rare because spontaneous mutations are rare. Individuals with early onset morbid obesity and severe disease states are unlikely to have children and the gene variant associated with their morbid obesity is less likely to be passed on to the next generation.

A further example of monogenic obesity relates back to material from Section 5.5, in which the role of αMSH in cells of the arcuate nucleus was described.

◆ What effect would you expect to observe in individuals who either were unable to make αMSH or, alternatively, could not synthesise the receptors for αMSH?

◆ In either case, they should suffer from early onset morbid obesity that is similar to that seen in individuals unable to manufacture leptin or its receptors.

In fact, this is exactly what clinicians have observed. An inability to make the receptor for αMSH, which is known as the **MC$_4$ receptor**, is thought to be the single most common form of monogenic obesity, present in about 6% of morbidly obese children. Remember though that such patients are extremely rare and represent only a tiny fraction of the total number of obese people worldwide.

Other cases of morbid obesity may arise as part of more complex genetic disorders. For example, Prader–Willi syndrome (Vignette 2.1) results from the loss of a large section of one chromosome that contains a number of different genes. It is associated with disorders of movement and cognitive impairment as well as overeating, leading to early onset morbid obesity that is very hard for parents to control. It occurs with a frequency of about 1 in 15 000–20 000 of all births. There are several similar disorders that are also associated with morbid obesity and, taken together, they are more common than monogenic obesity. However, they are also relatively rare causes of obesity.

Evidence for a genetic influence on obesity that extends beyond these relatively rare (but severe) cases comes from several sources including family, adoption and twin studies. Family studies look at the relationship between the BMI of family members. They show, for example, that the children of obese parents are very much more likely to be obese when they become adult. This was pointed out as early as 1923 in the scientific literature when Davenport wrote that this similarity 'suggests that there is a genetical difference in different families in the economy of fattening' – and hence the quotation that began this chapter (Davenport, 1923).

◆ In what different ways, using modern scientific concepts about genes and environment, might such family data be accounted for?

◆ (a) Family members are likely to share many of the same gene variants by inheritance. If certain gene variants predispose individuals to obesity, then this could account for the family resemblance.

(b) Family members, because of their close social relationship, tend to share the same environment, including diet, attitudes to food consumption and patterns of exercise. This shared environment could also account for the correlation of BMI within families.

Of course, these two explanations are not mutually exclusive; both effects are important and they interact with each other. However, it would be valuable to have a clearer idea of how large the genetic influence might be. In the mid-1980s, a series of studies of adopted individuals and their families was published by Albert Stunkard and his colleagues. They used Danish registers that included details of every adoption from the 1920s onwards and allowed the researchers to trace adopted individuals, together with these individuals' adoptive parents, biological parents, full siblings and half siblings.

In one study (Sørensen et al., 1989), they used a sample of adopted individuals, varying from lean to obese, to show that the adopted individual's BMI was positively correlated with the BMI of their biological parents (who they had not lived with) but was not correlated with the BMI of the adoptive parents (who they had lived with). In addition (Figure 6.2), the BMI of the adopted individual was positively correlated with that of their full siblings, from whom they had been separated since very early in life. Their BMI was also correlated, but to a smaller extent, with the BMI of their separated half siblings. An important conclusion from these adoption studies was that the genetic contribution to BMI is found right across the range of BMIs, from lean to obese.

> Siblings are brothers or sisters. Full siblings have the same mother and father; half siblings have one parent in common.

Research using twins has also helped to reveal the extent of the genetic basis of body weight control. For example, Wardle and colleagues (2008) used data collected in 2005 from a large sample of UK twins of about 10 years of age. They showed that the correlation in both BMI and waist circumference was very high in monozygotic twins but much lower in *dizygotic* (non-identical) twins. This allowed them to determine the relative importance of genetic and environmental factors in determining differences in body weight *when those individuals are living in the same environment*. The investigators concluded that as much as 77% of this variability was genetic. The remaining environmental effects were divided roughly equally between those that were shared between the twin pairs (e.g. the family environment) and those that were not shared in this way. This conclusion is very similar to that reached from studies on adults published some 30 years earlier: Wardle's study confirms these earlier findings. In addition, this study suggests that the genetic influence on obesity is apparent prior to puberty, and that it remains important in the modern obesogenic environment in which these children were born.

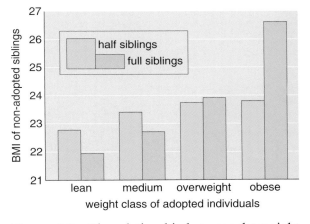

Figure 6.2 The relationship between the weight class of adopted individuals and the BMI of their siblings. The adopted individuals had been separated from their siblings at a very early age.

◆ It might be suggested that this, and similar studies, indicate that environmental influences on obesity are unimportant. Why is this view incorrect?

◆ Studies of this type investigate the reasons for *differences* between individuals in their body weight. An obesogenic environment may lead to *all* individuals being very much heavier than they would otherwise be, although the differences in weight *between* them might be relatively unchanged.

In another experimental study, 12 adult identical-twin pairs were observed over a period of 100 days in which they ate a diet that provided more energy than they required; in another, 7 adult identical-twin pairs were observed over a period of 93 days in which they ate a restricted diet that, together with an imposed exercise programme, placed them in energy deficit. The results from these two experiments are shown in Figure 6.3.

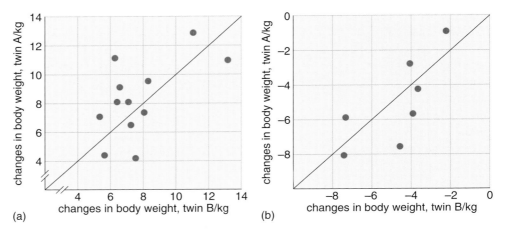

(a)   (b)

Figure 6.3 Weight changes in identical adult twins who spent (a) 100 days on a positive energy balance diet (12 twin pairs) and (b) 93 days on a negative energy balance diet (7 twin pairs). Each dot represents one twin pair. The diagonal line is where all the dots should be if the changes in weight were exactly the same within the twin pairs.

◆ Look at Figure 6.3a. These people all ate the same excess of calories; has it led to them all gaining similar amounts of weight? (Hint: Compare the dot furthest to the left with the one furthest to the right.)

◆ No. The dot furthest to the left shows that twin A puts on just over 7 kg whereas their twin B puts on just over 5 kg. The dot furthest to the right shows that this twin A puts on about 11 kg, whereas their twin B puts on just over 13 kg. So the 'right-side twins' have put on almost twice as much weight as the 'left-side twins'. Another way of putting this is that there is a significant correlation between the weight put on within the twin pairs.

A similar situation is seen in Figure 6.3b, in which the twins are losing weight. Again, there was substantial variation *between* twin pairs in their weight change. This suggests that genetically different individuals show great variation in their response either to an obesogenic environment or to an environment where energy deficit is likely.

So the two important results are:

• There was substantial variation between twin pairs in their weight change.

• Weight change within twin pairs was significantly correlated.

Family, adoption and twin studies suggest a substantial and very widespread genetic contribution to individual differences in BMI. The pattern of these results is not consistent with an effect that is produced by variants of just one or two genes. Instead small effects of many gene variants add together to produce genetic susceptibility to obesity.

In recent years, our knowledge of the human **genome** has increased greatly. The human genome is the complete set of human genes (numbering around 25 000) and their possible variants (of which more than a million are known). Using the technique known as *genome-wide association*, it is possible to look for an association of particular gene variants with an increased likelihood of obesity, measured by BMI or waist circumference. Large populations of individuals can be studied without having to specify the gene(s) of interest in advance, and relatively small effects on BMI can be detected. Variants of a gene known as *FTO* (fat mass and obesity-associated gene) have been shown to be associated with increased BMI in Caucasian populations, but not African or Asian populations (Loos and Bouchard, 2008). These *FTO* variants are very common among Europeans, with as much as 63% of the populations possessing at least one variant predisposing to obesity. Possession of one copy of such a variant predicts an increase in body weight of about 1.5 kg, and possession of two copies predicts just over double this amount. Although this is only a small increase, the widespread distribution of the variants of this gene means that they may explain about 1% of the total genetic influence on body mass in Caucasian populations. The function of this gene is not known, although it is expressed in some of the areas of the hypothalamus and hindbrain implicated in the control of eating behaviour (Section 5.5.1). This gene is likely to be the first of many which will be shown to be implicated in obesity using such techniques.

6.3   Environment and obesity

Although there is a substantial genetic contribution to individual differences in BMI and therefore obesity, the genetic contribution has not suddenly changed and so cannot provide an explanation for the very recent and substantial increase in levels of obesity seen in many human populations (Box 6.1).

Box 6.1 Rate of genetic change

How do we know that the genetic contribution to obesity has not suddenly changed in the last few decades? There are two important reasons underlying this statement. First, new variants of a particular gene can only arise by mutation, and spontaneous mutations are rare, only occurring in a small proportion of individuals in the population in any one generation. The mutation must occur in cells that give rise to eggs or sperm if it is to be passed on. Second, even when new gene variants do arise, it takes many generations for them to spread through the population. Any new gene variant in one person may only be passed on to one or two people, i.e. some of their children. Those children in turn may pass it on to some of their children, and so on. Only after many repeats of this process, i.e. after many generations, could the new gene variant come to be present in a sizeable proportion of the population. So both the rate of spontaneous mutation and the rate at which new gene variants pass through the population are slow. Recent changes in the prevalence of obesity have occurred in less than two generations, which is too fast for this kind of change.

Chapters 1, 2 and 5 have mentioned that there have been a number of changes in the environment, especially in Western societies, that strongly encourage the development of obesity. Many studies have shown how strongly these environmental variables correlate with increases in childhood obesity. For example, the increase in childhood obesity began between 1980 and 1988 and has become more pronounced in subsequent decades. At this time, high-calorie soft drinks and energy-dense foods were increasingly advertised to and consumed by children. In addition, the availability of television and video games was increasing and changes in the built environment were perceived as making it less safe for children to walk or cycle to school (see Figure 2.9) or to play unsupervised outside the home. Some of these changes in eating behaviour and exercise patterns may have been partly driven by increased numbers of parents who were working full-time. These are some of the factors that make up the obesogenic environment in which many children are now living.

Although the numbers of obese and overweight children have increased greatly since the early 1980s, children of healthy weight do still remain in the majority. Rates of overweight and obesity in adults are even higher, such that the majority now fall into the overweight and obese category in both the USA and some European countries (see Figures 1.1 and 1.2). But even for adults, there is a substantial minority comprising individuals who, despite their exposure to at least some of the components of an obesogenic environment, remain lean. The next section of this chapter shows how these individual differences might be understood by considering the complex nature of gene–environment interactions.

6.4   Gene–environment interactions and obesity

The gene for a protein known as PPAR-gamma (PPARγ) interacts with the environment to influence body weight in an interesting way. The gene itself is referred to as *PPARG* and is expressed in adipocytes. It is critical to the early development of adipocytes as well as to the regulation of fatty acid synthesis and release in the mature cell. *PPARG* may also be expressed in other cell types with different functions. The expression of *PPARG* is high in morbidly obese people but low in those who are lean. The function of the protein PPARγ is to act as a sensor for the levels of fatty acids within the cell and to regulate the expression of other genes that control metabolism in these cells. PPARγ could therefore be a critical molecule in the development of obesity.

A number of gene variants of *PPARG* have been described and one of these (*PPARG* Pro12Ala) has been associated with resistance to the development of diabetes. However, some studies have reported that possession of this gene variant is also associated with increased BMI, whereas others have reported lowered BMI or no significant effect.

Part of the reason for the discrepancies between studies in relation to the effect of the Pro12Ala variant of *PPARG* on BMI may be that it depends on the diet.

Genes are represented by taking the abbreviation for the protein and printing it in italicised capital letters.

Gene variants are often described in terms of the changed amino acid in the resulting protein. Pro12Ala means that the amino acid *alanine* has been substituted for *proline* at position 12 in the amino acid chain that forms the protein.

◆ Look at Figure 6.4 and describe the effect of diet on BMI.

◆ Individuals with the Pro12Ala variant of *PPARG* who typically consume a diet that is relatively high in polyunsaturated fat tend to have a lower BMI (of about 24.7) than those whose diet is relatively high in saturated fat (BMI of about 26.9). Although this is not a large difference in BMI, for some individuals it may make the difference between being of healthy weight and being overweight.

This is an example of a **gene–environment interaction**: such interactions arise when the effect of a gene variant is dependent on the particular environmental conditions. It is likely that these types of gene–environment interactions occur frequently.

Another gene with a number of common variants is *ADRB3*. This gene codes for a receptor that mediates some of the effects of the neurotransmitter **noradrenalin** on BMR (see Box 6.2). Noradrenalin release increases BMR. An important part of this action occurs through the effect of noradrenalin on adipocytes. Possession of the Trp64Arg variant of *ADRB3* – found in up to 25% of people, depending on the particular population that is studied – is associated with a less effective response to the neurotransmitter, and it predisposes towards increased BMI and visceral obesity. However, this effect is also subject to gene–environment interaction. BMI is increased in people possessing the Trp64Arg variant who have a sedentary lifestyle. People who have an active lifestyle, by contrast, are not at increased risk of obesity if they possess the Trp64Arg variant. Exactly the same relationship between physical activity and variants of *FTO* (Section 6.2) has been demonstrated in a large population of middle-aged Danish men. Individuals with two identical variants of the obesity-predisposing variants of *FTO* only showed an increased BMI if they also had an inactive lifestyle.

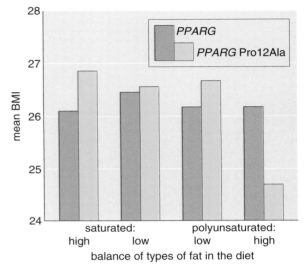

Figure 6.4 An example of a gene–environment interaction: the effect on BMI of two gene variants of *PPARG* (*PPARG* and *PPARG* Pro12Ala) depends on how diets vary in the balance of two types of fat.

Box 6.2 The many faces of noradrenalin

Noradrenalin – also known as norepinephrine – stimulates the body in three ways. One way is as a neurotransmitter in the brain. The second way is also as a neurotransmitter, but within the nervous system elsewhere in the body. It is released from a special set of nerves that variously link the spinal cord to the heart, gut, lungs, mouth and adipose tissue. The third, slower and more enduring way, is as a hormone via the bloodstream; in this case, it is released into the blood from the adrenal glands that lie just above the kidneys. There are a number of noradrenalin receptors, referred to as alpha- or beta-adrenergic receptors. Of particular interest are the beta3 (β_3) receptors, both because they are the main receptor subtype found on adipose cells and because they differ in sensitivity between adipose depots. Adipose tissue stimulated by noradrenalin produces glycerol and free fatty acids by lipolysis; the glycerol and FFAs are whisked away to where they are needed by the increased blood flow, enabling increased metabolic activity and raised BMR.

The *combined* effects of these variants of *PPARG* and *ADRB3* on BMI are unexpected, providing an example of **gene–gene interaction**. A study of obese and healthy weight children from northern Spain demonstrated that those with the Pro12Ala variant of *PPARG* had a raised BMI, but only when they also had the Trp64Arg variant of *ADRB3*. This effect is shown graphically in Figure 6.5 (fourth column). By contrast, those children with the most common variant of *PPARG* actually showed a reduced BMI compared with the norm if they possessed the Trp64Arg variant of *ADRB3* (Figure 6.5, third column) (Ochoa et al., 2004).

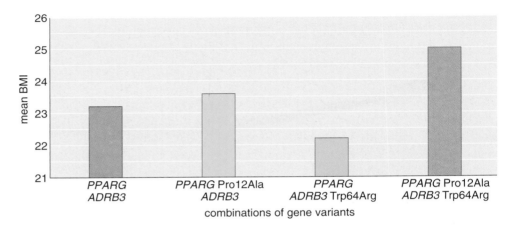

Figure 6.5 An example of a gene–gene interaction: the effect on BMI of gene variants of *PPARG* and *ADRB3*.

In this section we have looked at examples of the way in which common variants of genes such as *FTO, PPARG* and *ADRB3* can affect body weight. Several very important general points have emerged from these examples:

- The effects of these gene variants tend to be small and only explain a small part of the total variation in BMI that is present between individuals. However, considerable numbers of people may be affected because the gene variants are often common, and the effect on these individuals may be to move them from a healthy weight to overweight. The overall effect on the incidence of obesity within a population may therefore be very much greater than for a mutation, such as that in the leptin gene, which causes severe obesity but is extremely rare.

- The effects of combinations of variants for different genes may be complex and unexpected. Study of these *gene–gene* interactions and their influence on obesity is, in 2008, just beginning.

- In the same way, a particular gene variant may interact with environmental factors in complex and unexpected ways. This is known as *gene–environment* interaction.

- Here we have examined the effects of variants of just a small number of genes on body weight. However, it may be that body weight and adiposity are influenced by many different genes – perhaps as many as several hundred, each with several variants, although these genes will not have body weight control as their primary function. As a consequence, there is likely to be great variation in the possible interactions between genes and environment influencing body weight.

This chapter began with a brief description of the particular susceptibility of Pima Indians to the modern obesogenic environment. It is likely that this and at least some of the other ethnic differences in the prevalence of obesity are partly influenced by genetic variation. In the case of the Pima Indians, there have been a number of studies that have looked for gene variants that are associated with the occurrence of obesity. One straightforward result is that monogenic obesity is no more common in Pima Indians than in other populations (Ma et al., 2004). For example, a study of 300 morbidly obese Pima Indians (BMI of 45 and over) only found 13 instances of MC_4 receptor mutations, i.e. just over 4%, which is similar to the proportion observed in other groups of adults with this degree of obesity. Other studies have looked at variants of other candidate genes that might be associated with the raised susceptibility to obesity in Pima Indians. The majority of studies have not found any. For example, Section 4.6.2 mentioned that a reduced secretion of adiponectin is associated with obesity. There are several gene variants of adiponectin, but their occurrence is not associated with the degree of obesity in individual Pima Indians, although in a different study there was a positive association reported in a population of obese German people. Other studies have taken a different approach and measured the expression of a very wide range of genes in adipose tissue taken from this group of individuals. One important finding is a raised expression of genes mediating inflammatory responses, which is consistent with a raised secretion of TNF-α and IL-6 by adipocytes (Section 4.6.2).

A candidate gene in this context is one that codes for a protein, known on the basis of prior evidence, to be involved in some aspect of body weight control. Variants of such a gene might therefore be expected to be associated with variation in BMI or some other measure of obesity.

◆ Is the raised expression of one or more genes in an obese individual necessarily responsible for that individual's obesity?

◆ No. Although it is possible that raised expression of the gene occurred prior to obesity and was involved in promoting weight gain, it is as likely that the increased expression is a response to weight gain and increased fat deposition.

Gender also has a significant influence on obesity. Obesity rates are higher in women than in men (Section 1.6.1), although the consequences for ill health are reduced in women because of the different distribution of body fat. Body fat in the area of breasts, buttocks and thighs is much less likely to result in illness than high levels of visceral fat. At a genetic level, these differences arise, in part, as a consequence of the mechanism of sex determination. Early in development, a gene known as *SRY* is expressed and results in a male reproductive system developing. In its absence, a female reproductive system will develop. The consequence of the development of these different reproductive systems is that, especially after puberty, the hormones circulating in the bloodstream are markedly different in men and women. It is likely that the different circulating levels of sex hormones such as testosterone and oestrogen and the distribution of their receptors are powerful influences on the way in which body fat is laid down in a gender-specific manner. Testosterone is an *anabolic* hormone that promotes muscle growth, but lowered levels of testosterone (characteristic of ageing individuals) are especially associated with fat deposition in the trunk and abdomen and with visceral obesity.

There is an interesting comparison to be made between studies investigating the origin of ethnic and of gender differences in obesity. In the case of ethnicity, the emphasis has been on finding different clusters of gene variants that may affect a person's susceptibility to the obesogenic environment. In the case of gender, many studies focus on how gene expression is affected by events early in development. A brother and sister may share many of the gene variants that predispose to obesity. But the expression of these gene variants might be quite different depending on whether a single gene (*SRY*) happened to be expressed soon after fertilisation; i.e. depending on whether they were the male or female sibling. A person's current appearance and health may easily be affected by the expression of genes very early in life that influence the expression of other genes right through adult life.

In summary, there is considerable variation from one person to another in their genetic susceptibility to obesity. That variation arose over many years and will have been one part of the way in which the human species has been able to survive periods of famine in its evolutionary past. The legacy of that variation is for some people, and perhaps the majority, to be more susceptible to our modern obesogenic environment than others. As a consequence, they become obese. However, their obesity happens, not because they have a set of genes that *inevitably* lead to obesity, but because they are born into an obesogenic environment to which they are susceptible. We, both as individuals and as members of society, can, if we wish, make substantial changes to that environment and hence reduce the challenge of obesity that we face – a theme that is explored in Chapter 10.

The next section of this chapter focuses on the early stages of the human life cycle, when the challenge of obesity may be especially high.

6.5　Factors affecting childhood obesity

6.5.1　Early life

Gestation, the period of human pregnancy, sets the stage for a baby's start in life and may also contribute to their weight in the longer term. A baby's birth weight is easy to measure and has been found to be positively associated with later BMI. However, the relationship is not straightforward. One complication is that pregnancies differ in length, so the 'gestational age' at which babies are born is different; any differences in weight may simply reflect their different gestational ages. Any attempt to link birth weight with later BMI must take this into account.

A second complication is a phenomenon known as catch-up growth.

Undernutrition while in the womb leads to adaptive changes in physiology that prepare an individual for a life of poor nutrition. However, if poor nutrition prior to birth is followed by 'unexpected' good nutrition, rapid catch-up growth after birth tends to take place. Catch-up growth is not just relevant to historic times or to developing countries in which poor nutrition might be more prevalent. A study by Ong and colleagues (2000) put the incidence of catch-up growth at 30% at the end of the 1990s in the UK. Those infants experiencing catch-up growth have a significantly higher percentage of body fat at 5 years old (17.2%, compared with 15.8% for those infants growing at a normal rate). The difference in fat storage continues into adulthood. A long-term cohort study of growth and development in children (Demerath et al., 2007) found that babies with such catch-up growth tended to have greater fat mass and a higher percentage of body fat as middle-aged adults than those who experienced normal growth as babies (Table 6.1; Demerath et al., 2007). The trend was evident in men and women but was statistically significant only in men.

Table 6.1 Body fat measures of adults 45 years after showing normal or catch-up growth in infancy.

	Men			Women		
	normal growth	catch-up growth	trend	normal growth	catch-up growth	trend
fat-free mass/kg	67.0	68.8	ns[a]	48.1	48.6	ns
fat mass/kg	20.0	23.8	sig[b]	26.2	28.6	ns
body fat/%	22.8	25.4	sig	34.4	36.5	ns

[a]ns, difference not statistically significant
[b]sig, difference statistically significant

Infancy is obviously an important time for appropriate nutrition because rapid development of the brain and body is taking place. Breastfeeding of babies is the most appropriate nutrition for their needs, although feeding with artificial formula milk is sometimes required if breastfeeding cannot be established or maintained. Exclusive breastfeeding of babies is recommended by the WHO for a *minimum* of the first 6 months of a baby's life. The baby benefits because breast milk provides all nutrients plus protection against illnesses, infections and, it is believed, later development of obesity. This protection may, in part, be mediated by the baby having the opportunity to learn when to terminate a meal. The nutrient value of breast milk decreases as the feed progresses and the infant chooses when to stop feeding. By contrast, nutrients are uniformly distributed in formula milk and most parents will encourage the infant to take all of the feed. Additionally, formula-fed babies are often weaned onto food earlier (at around 16 weeks), leading to altered growth patterns and greater weight gain by the time they are a year old. It has been estimated that some 13 000 cases of childhood obesity could be avoided in England and Wales over a period of 7–9 years if all 600 000 babies born in 2002 were breastfed for a minimum of 3 months (Akobeng and Heller, 2007).

Satisfactory growth is used as an indicator of whether energy needs are being met. Monitoring of development has traditionally placed emphasis on weight gain, but this concern stemmed from historic times in which babies or children were more in danger of being *underweight* than overweight (Section 1.6.4). In the first few months of life, the energy requirement for growth relative to maintenance is larger than at all other times, hence the energy requirements of infants (Table 6.2) are different from adults.

Table 6.2 Estimated average requirements (EAR) for energy in the early years. Values are based on COMA data published by the Department of Health (1991).

Age	EAR/kcal d^{-1}	
	males	females
0–3 months	545	515
4–6 months	690	645
7–9 months	825	765
10–12 months	920	865
1–3 years	1230	1165

◆ Use Table 6.2 to compare the differences in energy requirements of male and female infants. Describe these differences.

◆ Male infants have higher total energy requirements (measured in kilocalories per day) than females.

◆ Why might full-fat milk be advised until at least 2 years of age?

◆ Dietary fat is an energy-dense food and so provides an important source of energy for young, growing infants.

This total energy requirement increases with age for both sexes. However, energy requirements adjusted for body weight decrease between 1 and 6 months of age (males: decrease from 113 to 81 kcal kg^{-1}; females: from 107 to 81 kcal kg^{-1}).

◆ How can the total energy requirement increase while over the same time period the energy requirement adjusted for body weight decreases?

◆ The reason is that the child is growing. A child growing from 8 to 16 kg would have to double their energy intake over the same period to keep a constant energy intake per kilogram (i.e. adjusted for weight). If they increase their intake by one and a half times, for example, their energy intake per kilogram would fall.

Energy deposition (storage) as a percentage of total energy requirements also goes down from 40% at 1 month to 3% at a year old. After this age, most children can walk and a lot of their energy is then used up, developing and using muscle tissue, rather than being stored.

◆ The energy requirement per kilogram of an infant is about 100 kcal kg^{-1} d^{-1}. How much larger is this value than the energy requirement per kg of an adult? (Use the value for sitting at rest from Table 2.2 for the adult.)

◆ A 70 kg adult requires 100 kcal h^{-1}, which means that their daily energy requirement is 2400 kcal d^{-1}. To find the adult's energy requirement per kg, divide this value by their weight:

$$\frac{2\,400 \text{ kcal d}^{-1}}{70 \text{ kg}} = 34.3 \text{ kcal kg}^{-1} \text{ d}^{-1}$$

So the infant's energy requirement per kg, at 100 kcal kg^{-1} d^{-1}, is about 3 times greater.

6.5.2 Childhood

Weaning onto solid foods from milk and the subsequent development of child feeding behaviour are seen by health care professionals as good timepoints at which to influence future healthy eating behaviour. It has been suggested that, once they reach school age, children could already be on the path to overweight and obesity. Parents, as the providers, enforcers and role models are seen to have the major influence on child eating behaviour and are often the focus of public health initiatives. For example, the policy document from the UK Government 'Healthy weight, healthy lives: a cross-government strategy for England' (Cross-Government Obesity Unit, 2008) states that 'because parents and parental behaviour has such a strong influence on child behaviour, excess weight problems in children can only be tackled in concert with tackling them in the whole family and society more broadly'. Such initiatives promote awareness about what choices can be made and provide information (e.g. via food labelling) about how to make those choices. They also occasionally offer specific guidance to parents, such as delaying weaning until at least 6 months old. The topic of public health policy is discussed further in Chapter 10.

A distinction can be made between *what* parents feed children and *how* they go about feeding their children (i.e. the rules and customs associated with food within the home), although clearly both are important. What parents feed children is to do with food choices, the quality, quantity and frequency of meals, the availability of snacks, the balance of foods and their energy densities.

Portion size is one of these food choices (Section 4.3.1). Fisher (2007) investigated how 44 5- to 6-year-old children responded when presented with slightly different meals. The main course (macaroni cheese) varied in quantity on different occasions, whereas other components (milk, biscuits, carrots and corn) were fixed. One main course was an age-appropriate 250 g of normal energy density (1.42 kcal g^{-1}); another was twice the size at 500 g. Each child received each meal type on a different occasion. None of the meals was energy-limiting and many children did not consume all of the age-appropriate meal. The energy intake of the children was a significant 15% higher with the larger portion size. Energy intake from other foods at the meal did not vary. A similar study found that 5- to 6-year-old children consumed 34% more energy when served a larger, more energy-dense (1.8 kcal g^{-1}) main course.

Another food choice is the provision of sweetened drinks. One Canadian study looked at the prevalence of overweight in preschool children in relation to their consumption of sugar-sweetened drinks between meals; i.e. they compared children who either did or did not consume these drinks between meals. Dubois and colleagues (2007) found that those children who consumed 4 or more sugar-sweetened drinks between meals per week were significantly more likely to be overweight between the ages of 2.5 and 4.5 years than those who consumed no such drinks between meals (Figure 6.6).

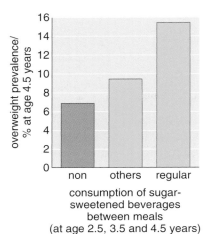

Figure 6.6 Percentage of overweight Canadian children at 4.5 years old in each of three categories, based on the frequency of their sugar-sweetened beverage consumption between meals. Regular: consume 4 or more such drinks between meals per week; non: never consume such drinks between meals; others: consume between 0 and 4 such drinks between meals per week.

These results reinforce the message that children are susceptible to consume more energy than they need. The converse is also true: minor changes in food availability could see a reduction in excess energy consumption by children.

How parents feed children is also subject to research. There are customs and practices about meals and food that different families adopt. Several studies have involved evaluating the extent to which parents agree with statements about controlling feeding practices. They rank on a scale of 1 (strongly disagree) to 5 (strongly agree) statements such as 'I have to be sure that my child does not eat too many high-fat foods' and 'If I did not guide or regulate my child's eating, she would eat too many junk foods'; parents with lots of 5s are categorised as controlling. Some of these studies suggest that such controlling feeding practices may be associated with children eating more in the absence of hunger, i.e. snacking, and sometimes with increased weight (Faith et al., 2004). At present, evidence about other parental styles is inconsistent, so, although interesting, such studies and their findings should be treated with caution.

By 'junk' food, we mean energy-dense food of little nutritional value.

Parental knowledge of nutrition – and of parenting skills to implement that knowledge – are likely to play a role in influencing the long-term eating behaviour of children. These few examples reveal that both the development and the avoidance of childhood obesity is a very complex area. (Some of the longer-term schemes that are currently being trialled to encourage healthy eating in children and their families are considered in Chapter 8.)

School age: eating and activity

Older children make their own food choices, especially over what to eat at school (or on the journeys there and back!). There has been considerable government involvement in the UK in what children eat, following media debate about the quality of school food. However, food eaten at school is only a part of the energy intake of children. The number of food outlets has increased in recent years while the relative cost of food outside the home and school environments has fallen. Hence the restrictions on access to food that used to exist are far less evident.

◆ Food choices are only part of the equation. What other choices by parents might make childhood obesity more likely?

◆ One big influence is the choice of transport: the shift from foot to car transport (Figure 2.9) might lead to a reduction in regular activity.

The Avon Longitudinal Study of Parents and Children (ALSPAC) is following the health and living conditions of 14 000 children born between 1991 and 1992 in Avon, UK. One study monitored the physical activity levels of 5500 11-year-olds with a strapped-on monitor. It was a **cross-sectional study** as opposed to a longitudinal study (Section 1.4.1), which means that they compared activity and adiposity measures in different children at one time. In other words, they took a measurement *across* their population of children. There was a clear negative correlation between the amount of physical activity and the amount of body fat, as well as between the amount of physical activity and BMI: the greater the amount of activity, the lower the amount of body fat and the lower the BMI. The data also suggest that a difference of 15 minutes of daily moderate or vigorous physical activity reduced the chances of being obese by nearly half (Ness et al., 2007).

The prevalence of childhood obesity in England is about 6% and rising (Figure 6.7). This value is based on BMI thresholds for overweight and obese children (Figure 6.8) established by the International Obesity Taskforce. Although these thresholds are not universally accepted (Section 1.3.2), they are regularly used.

The trend is international. Across the EU, the prevalence of overweight in children is expected to increase from 30.4% in 2006 to 36.7% in 2010 (Table 6.3).

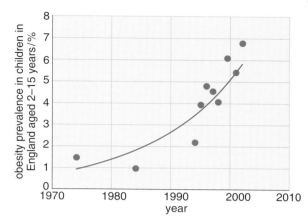

Figure 6.7   Prevalence of obesity in children in England. (Obesity defined using the International Obesity Taskforce BMI thresholds.)

Figure 6.8   Overweight children.

Table 6.3 Estimated levels of overweight and obesity among children aged 5–17.9 years in the EU.

Year	Prevalence/%	
	overweight or obese	obese
2006	30.4	7.1
2010	36.7	8.8

Activity 6.1   Treatments for overweight and obese children

Allow 10 minutes for this activity

Based on your reading of this section, what actions are available within families to treat overweight or obese children?

Now look at the comments on this activity at the end of this book.

Although most of the advice is generally applicable regardless of age, it is of note that weight *maintenance* is the aim in children.

◆ Why do you think weight maintenance is advised, instead of weight loss?

◆ Children have not yet reached their full height, and may be undergoing pubertal changes and changes in adiposity, so weight loss may be neither required nor advisable. Where children are obviously overweight, avoiding any unnecessary future gain is important, so weight should be maintained.

6.6 Pregnancy

Women often worry about their weight more than men (see Section 7.3.2). One reason may be that hormones such as oestrogen and progesterone also influence appetite. During a woman's menstrual cycle, there is a drop in food intake just before ovulation (release of an egg from the ovaries, ready for fertilisation). This occurs due to the influence of the rising levels of oestrogen. Soon after ovulation, a peak in food intake takes place due to the rising levels of progesterone that occur at this time. Within this cycle, these changes may relate to a difference of up to 80 kcal d^{-1}. However, once a woman becomes pregnant, far more extensive changes take place as the mother-to-be's BMI increases. There are also changes in appetite control and energy reserves and in the development and growth of the placenta and the baby (Figure 6.9).

An increased appetite (10–20% above normal) occurs early in pregnancy as well as after the birth, during **lactation** (milk production for breastfeeding; 20–25% above normal) and is maintained as long as breastfeeding is taking place. The term 'eating for two' is often used in relation to pregnancy but this certainly is a misconception as the second person (the baby) is much smaller than the first (the mother)!

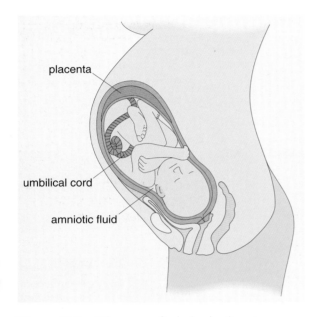

Figure 6.9   Diagram of a baby in the uterus, showing the placenta, umbilical cord and amniotic fluid.

The increase in BMR throughout pregnancy in the mother is due to an increase in body tissue mass (Table 6.4). These values have been calculated for an optimal pregnancy outcome with adequate milk production to support the baby's growth.

Table 6.4 Average additional energy requirements and body composition changes occurring during pregnancy.

Change during pregnancy	Overall change	First trimester[a] (birth to 3 months)	Second trimester (3–6 months)	Third trimester (6–9 months)
increase in body tissue mass	12 kg			
increase in fat storage	3.7 kg (range 3.1–4.4 kg)	8 g d^{-1}	26 g d^{-1}	−7 to 23 g d^{-1}
additional energy requirements	77 600 kcal[b]	90 kcal d^{-1}	290 kcal d^{-1}	460 kcal d^{-1}

[a]a trimester is a period of 3 months.
[b]for a healthy woman.

◆ Table 6.4 shows the increase in body tissue mass during pregnancy. What do you notice about the increases in fat storage?

◆ Fat storage increases during the first two trimesters of pregnancy, then there is a wide range from loss to gain in the third, final trimester.

This range in the third trimester is because of differences between individuals and their adiposity prior to the pregnancy.

A healthy total weight gain for pregnancy is approximately 12 kg (26 lb), and that includes the baby, placenta, amniotic fluid and increases in breast tissue, blood and body fluids and fat stores (Figure 6.10). It is worth noting that the increased fat depots formed during pregnancy are due to an enhanced response to insulin and tend to occur in the first and second trimesters. In the last trimester, which is the time within which the baby greatly increases in size, the mother is less likely to increase fat depots as she experiences a decreased response to the insulin she produces.

◆ What is the name given to the condition of reduced responsiveness to insulin?

◆ The condition is called insulin resistance (Section 4.5).

◆ What are the consequences of the mother's insulin resistance for the baby?

◆ The mother's insulin resistance will reduce the uptake of glucose from the blood into her cells, leaving more glucose circulating in the blood available to the baby and hence enhance its growth and development. This may be particularly important for the development of the brain.

Figure 6.10 Conspicuous changes in size and weight during pregnancy.

Insulin resistance in pregnancy is an adaptation to protect baby brains and, in a healthy pregnancy, the insulin resistance in the third trimester is transient; normal insulin sensitivity returns with the delivery of the baby. The rapid return to normal insulin sensitivity suggests a hormonal cause for the insulin resistance, although it remains unknown which – or which combination – of the many hormonal changes that occur during pregnancy might be responsible.

The additional energy requirement for lactation (compared with requirements when not pregnant and not lactating) is shown in Table 6.5 (overleaf), separated into the first 5 months when exclusive breastfeeding is taking place and the variable period after that in which partial breastfeeding may be taking place.

◆ Is the additional energy requirement during exclusive breastfeeding more than or less than during pregnancy?

◆ The extra energy requirement during the first 5 months of breastfeeding is 620 kcal d^{-1}, whereas the energy requirement during pregnancy range between 90 and 460 kcal d^{-1} (Table 6.4), depending on the trimester. The extra energy requirement during exclusive breastfeeding is therefore greater than at any time during pregnancy.

◆ To what extent are tissue reserves used to provide energy for lactation in well-nourished women?

◆ During the first 5 months of exclusive breastfeeding, 170 kcal d^{-1} of energy is mobilised from tissue reserves, which is equivalent to about 27% of additional requirements. After 5 months, only a negligible amount of energy is obtained from tissue reserves, hence the remainder is obtained from food.

Table 6.5 Energy requirements of lactation in well-nourished women exclusively breastfeeding for the first 5 months and partially breastfeeding thereafter.

	Exclusive breastfeeding	**Partial breastfeeding**
average quantity of milk produced	750 g d^{-1} (equivalent to ml d^{-1})	550 g d^{-1} (ml d^{-1})
additional energy requirement to produce milk	620 kcal d^{-1}	450 kcal d^{-1}
of which energy mobilised from tissue reserves	170 kcal d^{-1}	negligible
of which energy obtained from food	450 kcal d^{-1}	450 kcal d^{-1}

Weight gain and loss in relation to childbirth

Pregnancy is a time during which weight gain is not only normal, but is required. Fortunately for most women, that weight is readily lost soon after giving birth. Obstetricians suggest that women with higher BMIs should aim to gain less weight during pregnancy (Table 6.6). However, many women do gain more than recommended, with the weight accumulating in central (visceral) as opposed to peripheral adipose depots. A number of studies have documented that this excess weight gain and a failure to lose that weight within 6 months of birth is an identifiable predictor of long-term (within 10 years) obesity and associated comorbidities, such as diabetes and hypertension (see Chapter 7).

Table 6.6 Recommendations for total weight gain during pregnancy based on maternal BMI category.

BMI	Recommended weight gain during pregnancy/kg
Less than 18.5	12.5–18.0
18.5–24.9	11.5–16.0
25–29.9	7.0–11.5
30 and over	7.0

6.7 Epilogue

This overview of influences on early life stages is painted on a moving canvas. The relatively recent emergence of childhood obesity raises the question of what the health consequences are of long-term obesity. It also raises the question of whether childhood obesity will necessarily lead to more people falling into the category of the super obese, with a BMI of over 50. Unfortunately, before very long, we will find out.

6.8 Summary of Chapter 6

6.1 This chapter has emphasised some of the possible differences between individuals in relation to their weight gain potential, ranging from their gene variants and susceptibility to an obesogenic environment through to some critical life experiences.

6.2 Genes are inherited from parents and provide the instructions for proteins that make all the numerous hormones, hormone receptors, enzymes, etc. involved in metabolic pathways within the body.

6.3 Subtle differences in the same gene are called gene variants; the consequences of having a particular gene variant may depend on the environment in which it is expressed.

6.4 Where one defective gene product (e.g. leptin) results in obesity, the individual is said to have monogenic obesity.

6.5 Most obesity is due to a combination of many genes leading to a genetic susceptibility to obesity, often in combination with an obesogenic environment.

6.6 The incidence of obesity with known genetic causes is very low.

6.7 Mutations are too infrequent to be able to account for the sudden rise in the incidence of obesity since the 1980s. A more likely explanation is the way existing gene variants interact with the modern obesogenic environment.

6.8 The incidence of childhood obesity is rising. Many factors have been identified as contributing to this rise, including gestational, infant and childhood experiences.

6.9 Pregnancy should be a time of short-term weight changes, but these may lead to longer-term weight gain.

Learning outcomes for Chapter 6

LO 6.1 Define and use, or recognise definitions and applications of, each of the terms printed in **bold** in the text.

LO 6.2 Explain what is meant by gene–gene and gene–environment interactions in relation to the development of overweight and obesity.

LO 6.3 Outline the difference between monogenic obesity and genetic susceptibility to obesity.

LO 6.4 Describe factors during gestation, infancy and childhood that contribute to the risks of developing overweight and obesity in later life.

LO 6.5 Discuss how weight change can impact on pregnant women and their.

Self-assessment questions for Chapter 6

Question 6.1 (LO 6.1)

What is an MC_4 receptor and what does it do?

Question 6.2 (LO 6.1)

How might gene variants give rise to differences between individuals?

Question 6.3 (LO 6.2)

What is meant by the phrase gene–gene interaction?

Question 6.4 (LO 6.3)

Distinguish between monogenic obesity and genetic susceptibility to obesity.

Question 6.5 (LO 6.4)

A student decides to measure how much energy their 10-month-old son has eaten over a day and they arrive at a value of 800 kcal. Give one reason why they might worry. Give three reasons why they might not worry.

Question 6.6 (LO 6.5)

To what extent could the absence of breastfeeding contribute to a mother's weight gain following pregnancy?

CHAPTER 7

THE CONSEQUENCES OF OBESITY

Thou seest I have more flesh than another man, and therefore more frailty.

Falstaff to Henry IV in Act 4 of William Shakespeare's *Henry IV Part 1*

7.1 Introduction

By now, you should be convinced that obesity is not a cosmetic issue but a condition that, often very slowly, leads to physiological changes that have disabling consequences. In this chapter, you will study some of the medical conditions associated with obesity – the comorbidities first mentioned in Table 1.2 – in more detail, examining why obesity makes these conditions more likely to occur. From Chapter 6, you will be very aware that there is no single route to obesity and that, for most people, becoming obese is not inevitable. Many people who are not themselves overweight take that to mean that fat people have only themselves to blame. This moralistic attitude and the associated stigma that attaches to obesity is part of the environment in which the overweight child or adult has to live. There are therefore social consequences of being obese that affect all aspects of life, from school days to old age. Functioning in an unsympathetic environment has repercussions for the obese person's wellbeing, as you will learn in the second part of this chapter.

7.2 The medical consequences of obesity

7.2.1 Carrying too much weight

In Section 1.2 it was suggested that serious health problems created by obesity were related mainly to metabolic disturbances initiated by the accumulation of excess adipose tissue, whereas the additional weight associated with this adipose tissue was of less importance to health. In fact, you have probably realised by now that, because body systems do not work in isolation, this is too simple a view.

◆ One of the roles of the skeleton is to support the body (Figure 7.1a). While bearing in mind what has been said above, turn to Table 1.2 and say whether you think any of the health problems listed there might arise primarily as a result of carrying too much weight.

◆ Lower back pain might result from overloading the spine and osteoarthritis in knees might result from overweight.

Mobility is frequently a problem for those who are overweight or obese. When fat and lean tissues increase in mass, the skeleton, heart, lungs and other organs such as the kidney that have to 'service' this increased mass do not increase in size accordingly. Large bodies are mechanically inefficient, meaning that movements have to take account of the size of limbs and trunk; for example, arms cannot move in a direct line forwards and back when walking, but must be slightly raised in order to pass the enlarged trunk of the body. Not infrequently, this difficulty with movement leads to poor posture and subsequently to uneven weight falling on joints and damaging them – particularly the hips, knees and spine.

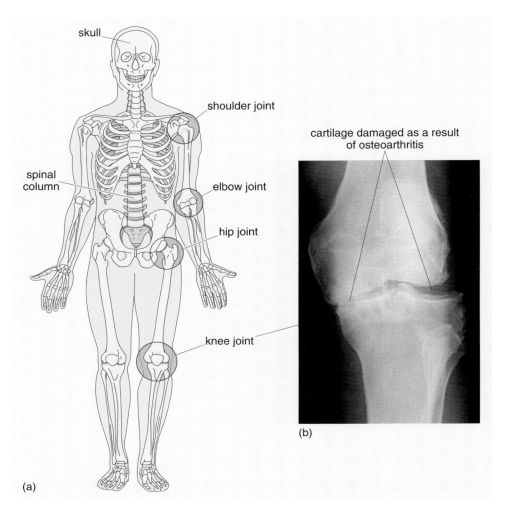

(a)

(b)

Figure 7.1 (a) The human skeleton. (b) A knee joint showing osteoarthritis. Note that the ends of the bones within the joint show uneven wear.

In osteoarthritis, the softer tissue (called cartilage) at the ends of bones (Figure 7.1b) or as the 'discs' between the bones of the spine (vertebrae) can be seen to have worn unevenly. The surrounding area is often inflamed. The ensuing pain discourages movement and hence there is a tendency for a vicious circle to develop, with greater weight gain leading to further reduction in mobility. One possible treatment is joint replacement surgery. However, some clinics do not offer joint surgery for those with a BMI of 40 and over because they do not have the necessary specialist equipment to cope with any complications that might arise. Anti-inflammatory drugs such as ibuprofen offer only limited pain relief, so it is unsurprising that numerous surveys on quality of life find that people who are obese and have osteoarthritis also suffer from depression. For those who do have joint surgery and have no post-operative complications, the outcome can be totally successful, irrespective of BMI.

Sleep apnoea – a condition in which someone temporarily stops breathing (**apnoea**) while asleep – is also primarily a consequence of the physical presence of fat, this time in the depots around the neck and the *viscera* (body organs). The increased amount of fat in the chest wall and abdomen in an obese individual causes a reduction in lung volume and alters the pattern of breathing. This is accentuated by lying flat. The *compliance* of the lungs is reduced by the increased mass of fat. Compliance refers to the ability of the lungs to stretch and thus alter the volume of air contained within them. Hence a reduction in compliance, together with a reduction in lung volume, will cause a reduction in the total volume of air which the lungs can hold, and in the amount of air that is inhaled and expelled with each breath.

During one of the phases of sleep, known as rapid eye movement (REM) sleep, muscles lose *tone* (i.e. they become 'floppy'). The airways can become obstructed (blocked) by loss of tone of the muscles which control the tongue. This occurs more frequently in obese individuals, particularly if the neck is thickened by excess fat. Breathing may be irregular and sleep apnoea may occur. Apnoea results in the blood carrying less oxygen and more carbon dioxide than normal, a situation detected by sensors that send the information to the brain's respiratory centre, initiating a sudden intake of air. This violent snore disturbs sleep. If apnoeic episodes are frequent, blood oxygen levels can fall very low. In some people this may lead to an irregular heart beat (cardiac arrhythmia), which can be life-threatening.

In its most severe form, individuals have prolonged episodes of sleep apnoea, causing frequent waking during the night and excessive sleepiness in the daytime. Eventually, low oxygen and high carbon dioxide levels in the blood may be persistent. If blood pressure is high (which it often is, for reasons to be explained in the next section), the pressure of blood passing through the lungs increases, causing the right side of the heart to beat less efficiently (right-sided cardiac failure). This is known as **obesity–hypoventilation syndrome** and can cause severe disability and death.

Although the description of sleep apnoea began by ascribing the problem to the presence of excess fat depots, the consequence is altered physiological function affecting both cardiovascular and respiratory systems. The next section looks at other aspects of cardiovascular functioning affected by obesity.

7.2.2 Cardiovascular conditions

As a person's weight increases, lean and fat mass increase, and the amount of oxygen required by the body therefore increases too. The total volume of blood in the body also increases in proportion to the increase in body weight. The rate at which the heart beats does not alter greatly; however, the volume of blood pumped out of the heart with each beat increases in order to meet the body's increased requirement for oxygen. Eventually, this will cause thickening of the wall of the largest chamber of the heart, the left ventricle (Figure 7.2). The mass of the left ventricle increases in proportion to a person's BMI. This is a point that we will return to when considering diets and exercise in Chapter 8.

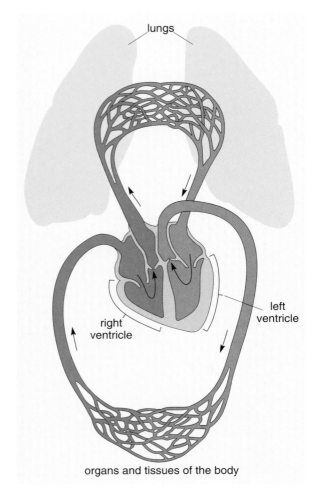

Figure 7.2 Schematic diagram showing the blood circulating through the heart for distribution to the lungs (from the right ventricle) and to the rest of the body (from the left ventricle), viewed as if facing the body. The arrows show the direction of blood flow.

◆ Look at Figure 7.2 and suggest why it is the left ventricle – not the right ventricle – that enlarges.

◆ The left ventricle pumps blood from the heart to all the body systems except the lungs, which are supplied by the right ventricle. Therefore the left ventricle has to exert more force than the right ventricle. As the body increases in size, the left ventricle enlarges so it is able to continue to fulfil this function.

Further changes in heart anatomy will occur as the increased volume of blood pumped by the heart causes an increase in the size of the cavity (interior) of the ventricles; this reduces their ability to contract. Hence the chambers of the heart are incompletely emptied with each heart beat and the heart is unable to generate blood flow sufficient to meet the demands of the body. This is a cause of oedema, which can occur in the legs and arms or in the lungs. Ultimately, the heart will fail.

The progression from structural changes of the heart to heart failure will be more rapid if there is also hypertension. Approximately 40% of obese individuals have hypertension (see Vignette 7.1). Blood pressure is the force exerted by blood on the walls of the blood vessels – this is determined by the force with which the heart contracts to pump blood through the blood vessels and by any resistance to this blood flow within the blood vessels. Hypertension normally occurs when there is increased *resistance* to the flow of blood through blood vessels. As a consequence, the heart has to work harder than it should to move blood around the body.

One reason why obesity increases the risk of developing hypertension is that obesity predisposes towards **atherosclerosis**, otherwise known as hardening of the arteries, in which fatty material accumulates within the blood vessel wall, leading to narrowed arteries and restriction of blood flow. In addition, raised insulin levels in insulin resistance (Section 4.5) cause the body to retain sodium (less is disposed of in the urine). To counteract the raised levels of sodium, the body retains water – a homeostatic mechanism to maintain an appropriate balance of this mineral in body fluids. However, the retention of water means that the volume of body fluids increases, increasing the blood pressure.

◆ Having read Vignette 7.1, what symptoms do you think hypertension has caused for Angus?

◆ Angus is not experiencing any symptoms of hypertension. He goes to see his doctor about something completely different and his doctor decides to measure his blood pressure while he is there.

Vignette 7.1   Introducing Angus

Angus is a 42-year-old solicitor who lives and works in London. He is 1.85 m tall and weighs 124 kg. He used to play rugby but had to stop two years ago due to a knee injury, and he has gained 20 kg since then. He enjoys eating out with his friends and is a very good cook; he particularly enjoys red meat and uses a lot of cream and oil in his cooking. He also considers himself to be a wine connoisseur and can't remember the last time he didn't have a drink in the evening; if he is socialising, he will drink at least a bottle of wine. He doesn't worry about his weight as many of his friends are of similar build.

Angus goes to see his local doctor because he has been getting a lot more indigestion recently and his doctor finds that his blood pressure is high. His doctor tells him he must lose weight and prescribes some tablets for his blood pressure. Angus doesn't know why he should lose weight or, indeed, how he should go about it. He thinks he eats pretty healthily and, although he considers taking exercise, his knee injury is a disincentive. His doctor didn't seem too worried about his blood pressure, so Angus doesn't worry about it either, and often forgets to take his tablets. At his follow-up appointment, his blood pressure is still high and he admits he often forgets his tablets.

Shortly after his second visit, Angus applies for life insurance and is astonished to be turned down on medical grounds. When he questions his doctor about this, he is told that it is due to his blood pressure and weight, and the fact that he has not followed his doctor's advice. Angus really cannot see what the fuss is about and decides to find a life insurance policy with an application form that he can complete himself.

Recall that Section 4.3.1 stated that insulin has numerous roles.

Hypertension does not cause any symptoms in the majority of cases. This can make it very difficult to treat because many people do not bother to follow advice when they feel completely well. However, as you have just read, if left untreated, hypertension increases the risk of developing heart disease. It also increases the risk of having a **stroke**, which can be caused by burst blood vessels in the brain. Strokes can also be caused by the brain's blood vessels being blocked by a **thrombus** (blood clot). A thrombus is particularly likely to be formed at a site of atherosclerosis, but is actually more dangerous if it breaks away because it may then reach smaller vessels and completely block them. If this occurs in arteries supplying the brain, it will result in a loss of blood supply to part of the brain. Tissues cannot survive without a blood supply bringing oxygen and nutrients, and so the 'starved' part of the brain will die. This may cause disability, such as paralysis or inability to speak, or, if it affects an area controlling a vital function such as breathing, death.

7.2.3 Diabetes mellitus

Diabetes mellitus is a condition in which the blood glucose level is higher than it should be, and if it remains that way, over time, it will cause numerous medical problems including cardiovascular diseases. In Western industrialised countries, diabetes affects about 4% of the population. Worldwide, about 90% of people with diabetes have type 2 or maturity onset diabetes and about 10% have type 1. In type 1 diabetes, the pancreas is unable to produce insulin, whereas in type 2 diabetes, insulin is produced, but either in insufficient amounts, or the target cells have a reduced response to it. Around 70–80% of cases of type 2 diabetes in adults can be attributed to the development of overweight and obesity (Section 1.5), and the number of overweight and obese children who develop type 2 diabetes is increasing and giving cause for concern. This was a disease that, until recent times, rarely occurred in anyone under the age of 40.

In all cases of type 2 diabetes, the pancreas initially continues to produce insulin. Insulin resistance develops when target tissues such as muscle and liver reduce their response to insulin and fail to remove glucose from the blood (Section 4.5). The level of glucose builds up in the bloodstream, resulting in hyperglycaemia.

◆ What effects do these high levels of glucose have on the insulin-resistant tissues?

◆ The liver responds as if it is fasting: gluconeogenesis is stimulated, which releases yet more glucose into the bloodstream. Similarly, in adipose tissue, fat reserves may be broken down, releasing fatty acids and glycerol into the bloodstream so that the liver can make yet more glucose.

◆ What is the effect of insulin on fatty acid metabolism?

◆ Normal fatty acid metabolism can take place. This means that, as long as the precursors (glycerol and fatty acids) are readily available (which they would be in the diet of someone who is well fed), TAG synthesis will occur, causing adipose depots to expand (see Figure 4.13).

◆ Insulin resistance can develop in anyone. Why do you suppose it is more likely to develop in someone who is overweight or obese?

◆ Individuals who are overweight or obese are likely to have been or to be eating more food and/or more frequently than non-obese people. Continued high levels of blood glucose provided from their diet means continual high levels of insulin in the blood. Receptors for insulin tend to be downregulated (not available for use; see Section 3.4.1) after prolonged exposure to high blood glucose levels, meaning that glucose is not taken up by target cells but remains in the bloodstream.

Additionally, replete adipocytes reduce production of adiponectin (which improves insulin sensitivity) and increase production of other adipokines, such as RBP4, which block the action of insulin (see Figure 4.16).

You might also recall that some of the signalling molecules produced by replete adipocytes are depot-specific – for example, visfatin is produced by visceral fat depots. Differences such as this may eventually increase our understanding of why visceral fat is associated with more adverse health outcomes than subcutaneous fat. It is already known that visceral fat depots respond more strongly than gluteofemoral depots (see Figure 4.15) to signals from the neurotransmitter noradrenalin.

◆ How does noradrenalin affect fat depots? (See Box 6.2.)

◆ Noradrenalin encourages lipolysis of fat stores.

Thus visceral fat releases more fatty acids, encouraging the pathway to hyperglycaemia.

Type 2 diabetes develops slowly and may be present for many years before a clinical diagnosis is made. In many cases, the condition is not recognised until pancreatic cells are damaged and produce little insulin, resulting in sudden weight loss (Section 4.5). Some people have few obvious symptoms of diabetes, and some pregnant women who are diagnosed with diabetes (known as **gestational diabetes**; see also Vignette 7.2) may have had undiagnosed diabetes prior to their pregnancy. In most cases, however, gestational diabetes is transitory and physiology returns to normal after the child is born (Section 6.5.2). Having diabetes for several years before a diagnosis is made can mean that complications of diabetes, including cardiovascular diseases, may already be present at the time of diagnosis. A current concern, with the rising prevalence of childhood obesity, is that insulin and glucose metabolism may be altered at a younger age, bringing forward the age at which complications will develop.

So why do we worry about diabetes? Diabetes significantly raises an individual's risk of developing various other conditions because prolonged periods of elevated glucose in the bloodstream causes damage to large and small blood vessels throughout the body as a consequence of glycation (Section 4.2). Damage to the large blood vessels can result in heart disease and strokes (as you have just read), and damage to the small blood vessels can cause disability. For example, damage to the blood vessels of the eyes (retinopathy) leads to blindness; diabetes is the leading cause of blindness in adults. Damage to nerves (neuropathy) can result in loss of sensation in the body extremities and lead to undetected damage such as ulcers (open sores), especially on the soles of the feet (Figure 7.3). Gangrene can ensue and it may be necessary to amputate part of the limb.

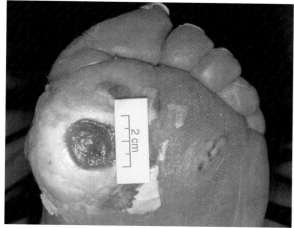

Figure 7.3   People with diabetes are at risk of developing foot ulcers such as this, which can take a long time to heal.

Vignette 7.2 Nazneen's pregnancy

Nazneen (see Vignette 1.2) is now 27 and is currently 26 weeks pregnant with her first pregnancy. Nazneen and Tariq had been trying to conceive for several years, so were delighted when they found out she was pregnant. Throughout the pregnancy, Nazneen has felt constantly hungry and her family reassure her that it is fine for her to eat what she wants as she is now eating for two. Before her pregnancy, she weighed 79 kg; she now weighs 102 kg and is 1.65 m tall. At a routine antenatal appointment, she is found to have glucose in her urine. Nazneen's midwife explains that she could have developed gestational diabetes and that she must have an oral glucose tolerance test (OGTT). Her midwife also warns Nazneen that her excess weight is putting her health and that of the baby at risk. She gives Nazneen some leaflets about healthy eating in pregnancy and arranges the appointment for the OGTT. Nazneen goes away feeling desperately worried and guilty that she may be putting her baby's health at risk.

Nazneen is given a 9 a.m. appointment for her test. She is asked to fast for 12 hours before the test and to bring a bottle of Lucozade® with her. The basic idea behind the OGTT is to load the bloodstream with 75 g of glucose and then see how rapidly the glucose is removed. If the system is performing properly, i.e. it has normal tolerance for glucose, blood glucose levels should have returned to normal within 2 hours of taking the test. In normal circumstances, skeletal muscle glucose uptake will account for the disposal of 70–90% of this oral glucose load, with adipose tissue accounting for 5–15%. This requires that both tissues are responsive to insulin.

The nurse takes a sample of Nazneen's blood and then asks her to drink her Lucozade. Nazneen is then asked to return 2 hours later and has another blood sample taken.

The following week, Nazneen returns to the surgery with Tariq for her results. She is told by the midwife that her fasting blood glucose level is 6.3 mmol l^{-1} and her 2 hour blood glucose level is 9.1 mmol l^{-1}. (Compare these values with those shown in Figure 4.4a.) Based on this result, she has **impaired glucose tolerance**. She does not have diabetes but is at risk of developing diabetes in the future. The midwife suggests that if Nazneen is able to lose weight, this will reduce insulin resistance, which will, in turn, reduce her risk of developing diabetes in the future.

Lucozade® is an energy drink containing a known concentration of glucose.

7.2.4 Dyslipidaemia

Technically, the term **dyslipidaemia** refers to any disturbance of blood lipid levels. By far the most common condition is **hyperlipidaemia**, when levels of lipids in the blood are higher than normal. Fatty acids and cholesterol are carried in the blood as **lipoproteins**, which are aggregates of lipids and proteins (Section 2.4.2). As shown in Figure 3.14, dietary fatty acids re-form as TAGs within gut cells and then aggregate with cholesterol and phospholipids, forming vesicles that enter the lymphatic system and then the blood system.

These vesicles are the largest and the lightest of the lipoproteins found in the blood because they contain much more fat than protein, and fat has the lower density. When they reach the liver, they may be reprocessed as very-low-density lipoproteins (VLDLs; smaller aggregates that also include cholesterol) and released back into the bloodstream. To get the TAGs from the blood into adipocytes for storage, they have to be broken into smaller units again. The adipocytes secrete a lipase that becomes anchored in the wall of the blood vessel, where it breaks down the VLDLs, freeing the fatty acids and glycerol (Figure 7.4). As the TAGs are removed, the proportion of protein and cholesterol in the VLDL increases, their density increases and they become **low-density lipoproteins (LDLs)**. Cholesterol has a number of vital roles (Section 2.4.2). So cells that require it take it up from the LDLs. Cells with surplus cholesterol export it to the liver as **high-density lipoproteins (HDLs)**. Obesity causes an increase in LDLs and a decrease in HDLs. A raised level of LDLs increases the risk of developing atherosclerosis because lipids leak from the LDLs and are deposited in the walls of blood vessels. In contrast, there is some evidence that HDLs can remove cholesterol from areas of blood vessels.

◆ Why might the lifestyle of an obese individual cause an increase in LDL cholesterol?

◆ One possible explanation is that an obese individual may have a diet which is higher in fat and cholesterol than a non-obese individual. The cholesterol taken into the body in food is likely to cause an increase in LDL levels.

All our cells can synthesise cholesterol, but usually not enough to meet their needs; the liver makes up the shortfall. If dietary cholesterol intake increases, then the liver cuts back on cholesterol synthesis, so cholesterol intake only becomes a problem if it is excessive or if there is some other metabolic dysfunction.

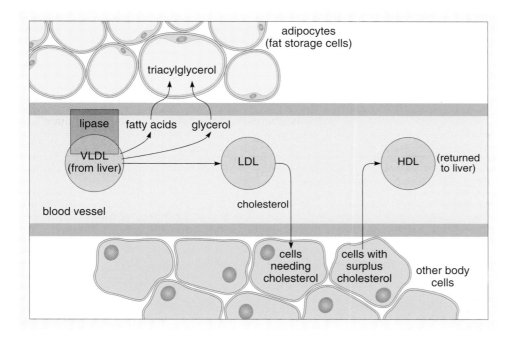

Figure 7.4 Schematic diagram showing lipoprotein and cholesterol transport.

7.2.5 Gall-bladder disease

Figure 3.1b shows the gall bladder. This is where bile, produced by the liver, is stored until its release during digestion. Bile is a watery solution containing cholesterol, fats, bile salts, proteins and bilirubin – a yellow-coloured waste product. Excess cholesterol in the bile may form hard deposits called gallstones and may also reduce the efficiency of the emptying of the gall bladder. There may be pain as gallstones pass through the bile duct. Very large stones may block the duct, resulting in persistent pain with fever and jaundice – yellow colouration of the skin. As shown in Table 1.2, obesity greatly increases the risk of gall-bladder disease.

7.2.6 Cancer

The relative risk of developing cancer when obese is much less than the risk of developing gall-bladder disease (Table 1.2).

◈ Far more media attention is given to the link between cancer and obesity than to the link between gall-bladder disease and obesity. What do you suppose are the reasons for this?

◆ Cancer has the reputation of being a 'killer'. The vast majority of people have heard of it; many know someone who has died of it; and many are themselves afraid of it.

You might also have suggested that cancer is the more common condition and is more difficult to treat. The prevalence of cancer – at around 4% of the population – is estimated to be more than 5 times higher than that of gall-bladder disease; treatment takes longer and can be very costly. Cancer Research UK estimates that in Europe 5% of cancers in women and 3% in men are caused by overweight and obesity, and that in the UK alone 12 000 people per year might avoid cancer if they maintained a healthy weight.

Cancer is a group of diseases characterised by the unregulated growth and spread of abnormal cells which, if not controlled, can result in death. Obesity increases the risk of certain cancers. One mechanism behind this may relate to insulin's ability to stimulate growth of cells. It is possible that the high levels of insulin seen in obesity due to insulin resistance could increase the risk of the uncontrolled growth of cells in certain tissues. Another mechanism, relevant to breast and uterine cancers, may relate to the fact that adipose tissue produces oestrogen, one of the hormones essential in regulating the female reproductive system but that, in excess, can stimulate abnormal cell growth.

The main types of cancer for which obesity is a risk factor are listed below.

• Breast cancer: excess oestrogen is a risk factor for the development of breast cancer, in both men and women. (Growths are also more difficult to spot in screening tests when the woman is obese.)

• Uterine cancer: increased oestrogen levels in obese women can cause thickening of the lining of the uterus, which can ultimately lead to cancer.

- Colon/rectum cancer (cancer of the bowel): this may be due to raised insulin levels in obese individuals or to a high level of saturated fats in the diet. The risk is also greatest for men with visceral adiposity.

- Pancreatic cancer: probably a consequence of hyperinsulinemia.

- Kidney cancer: probably a consequence of hyperinsulinemia.

7.2.7 Effects of obesity on fertility

Obesity is linked with impaired fertility in both men and women (see Vignette 7.2). The increased levels of circulating insulin seen in insulin resistance stimulates the body to produce a variety of hormones, including the sex hormones oestrogen and testosterone, at the same time reducing the production of their carrier protein, sex-hormone binding globulin (SHBG). Only hormones which are not bound to SHBG can act on target tissues. Thus a reduction in SHBG production exacerbates the increase in free hormone levels.

In women, the increase in free testosterone can adversely affect the functioning of the ovaries. It can also cause excessive facial and body hair. Although some obese women have normal ovarian function and fertility, this raised level of testosterone in the bloodstream can lead to a reduction in fertility by interfering with ovulation (the ovaries' ability to produce an egg for fertilisation). Leptin can inhibit ovarian function too, yet weight loss of as little as 5% can restore ovulation.

In men, raised levels of oestrogen have the effect of lowering levels of testosterone, and the result can affect external appearance (Figure 7.5). There is some evidence of reduced numbers (but not quality) of sperm, and there is also more reported erectile dysfunction in obese men than in men of healthy weight.

Figure 7.5 Obese men often appear to have breasts. As well as raised levels of oestrogen, obese older men can have looser skin and poor muscle tone, which contribute to their altered external appearance.

7.2.8 Complications in pregnancy and the newborn

From Table 7.1, you can see that being overweight or obese while pregnant can cause a number of problems. Obese women who do conceive are 3 times more likely than non-obese women to experience a miscarriage or have a baby

Table 7.1 Implications of excess adiposity on pregnancy in overweight and obese women.

| | Complications | |
Stage	Medical	Technical
early pregnancy	miscarriage; abnormalities in the developing baby	ultrasound examination: it is difficult/impossible to view the developing baby
antenatal	pre-eclampsia; gestational diabetes; deep vein thrombosis (DVT; blood clots in veins deep in the leg)	maternity scales have a maximum weight of 125 kg; blood pressure cuffs may be too small
labour	induction or caesarean section often necessary; baby getting 'stuck' part-way through delivery	delivery bed/operating table too small; fat absorbs anaesthetic unpredicatably; potential cardiovascular and breathing difficulties
at birth	haemorrhage (excessive bleeding); infection; DVT	

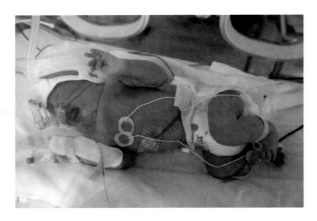

Figure 7.6   Baby born to an obese mother being monitored for heart defects.

with a deformity such as spina bifida. They are also twice as likely to have a baby with heart or other developmental defects (Figure 7.6). Increased infant mortality also occurs. Women with obesity have an increased potential to experience pregnancy complications: up to a threefold increase in risk of pre-eclampsia – a condition of acute hypertension and oedema – and a fourfold increase in risk of gestational diabetes. Obesity is a factor in 35% of maternal deaths in the UK.

As discussed in Section 1.5, it can be difficult to estimate the economic costs of overweight and obesity, but some French studies cited by Heslehurst and colleagues (2007) put the increased costs of prenatal care in overweight and obese women at between 5 and 16 times higher than women of healthy weight, and the percentage of babies admitted to intensive care at 3.5 times higher. Health care practitioners interviewed by Heslehurst and colleagues identified maternal obesity as having a number of impacts on the level of care required, including cost and equipment issues, and risk and complications for both mother and child – some of which related to existing maternal disease complications due to overweight/obesity. This study also identified the lack of information about psychological issues: how to inform pregnant overweight/obese women of the risks to themselves and their baby without causing undue concern and appearing to victimise them, especially when most of them are unaware of the health issues. Women generally are motivated to ensure the health of their children, so it is a key time to get important health messages across.

7.3   Psychosocial consequences of obesity

The requirement for tactful communication with an obese patient is not exclusive to pregnancy. This may partly explain why many doctors do not discuss a patient's weight unless the patient raises this as an issue. In one study, less than half of all GPs advised obese patients attending routine checkups to lose weight or discussed the potential implications of obesity on health (Galuska et al., 1999). Whereas some obese individuals are aware that their health is at risk because of their weight, many others do not realise this (see Vignette 7.1).

Obesity is a highly visible condition, so although an obese individual might not be aware of their health risks, they will be very aware of discriminatory and negative attitudes towards them.

7.3.1   Discrimination and negative attitudes

Discrimination is defined as bias in the treatment of a person or group which is unfair or unlawful. The extent of the discrimination that obese individuals face is being increasingly recognised. Puhl and Brownell (2001) looked at evidence of discrimination from a range of studies. Their conclusions relating to education, employment and health care are given below, together with the findings from some more recent studies.

Education

Discrimination begins early in life. Children as young as eight have been found to have negative views towards overweight and obese peers, with obese children often developing low self-esteem and feelings of guilt about their weight at a similar age. These obese children are teased and denigrated not only by peers but even by members of their immediate family. Negative feelings are also seen in teachers, as described by the following person seeking treatment for obesity (Puhl et al., 2005).

> When I was a child, I was sick and absent from school one day. The teacher taking attendance came across my name and said 'She must have stayed home to eat'. The other kids told me about this the next day.

A substantial proportion of teachers in one study believed obese individuals to be untidy, to be less likely to succeed at work and to be undesirable marriage partners; 28% of those questioned agreed that becoming obese is one of the worst things that could happen to a person.

Discrimination is also reported in entry to higher education. One study found that obese students – and, in particular, obese women – were significantly less likely to be offered a college/university place after interview, despite having similar academic qualifications and performance to non-obese peers. Distressingly, those who did gain admittance – again, particularly if they were women – received less financial support for their studies from their families than did thinner children.

Employment

A number of studies have been carried out to examine weight-based discrimination in the workplace. It is a difficult area to investigate as employers can claim that they have chosen to appoint an individual because of their qualifications or experience and thus deny discrimination. However, experimental studies have been designed in which participants are asked to evaluate the qualifications of different applicants for a job where some of the applicants are described or pictured as overweight or obese. In several studies of this kind, employers have been found to have more desire to employ non-obese applicants even if obese applicants have similar or better qualifications. In addition, obese individuals are more likely to be considered for telephone work rather than face-to-face interactions compared with non-obese people. Obese individuals attending interviews are more likely to be judged as less tidy, productive, ambitious and disciplined than their non-obese counterparts.

This preference for employing non-obese individuals may help explain why fewer obese individuals are hired in higher-ranking positions, and why obese individuals are less likely to gain promotion than their non-obese colleagues. It also accounts for a wage differential, with obese women likely to earn significantly less than non-obese women. The same wage difference does not occur in men. However, they are underrepresented in professional or managerial positions. Obese men are overrepresented in occupations related to transport, suggesting that they chose occupations where they feel their weight will not hinder their earning capacity.

In addition, there are various reports of employment being terminated because of an employee's weight. There are several instances of former employees taking their employers to court, claiming termination of employment was due to weight discrimination. However, the existence of such cases does not necessarily mean that weight discrimination leading to termination of employment has occurred – only that some employees *believe* they have been unfairly treated due to their weight.

Health care

In studies that go back to the 1960s, a consistent finding is that health care professionals share with the general population the negative perceptions towards obese people. A carefully designed questionnaire found that even professionals specialising in obesity (including research scientists, physicians and dietitians) have implicit negative attitudes towards obese individuals, believing them to be lazy, lacking in self-control and unsuccessful (Schwartz et al., 2003).

◈ What is meant by an 'implicit' attitude?

◆ An implicit attitude is an attitude of which you are unaware. (Likewise, implicit learning was discussed in Section 5.4.1.)

A US study (Foster et al., 2003) found that more than half of 620 doctors thought their obese patients were awkward, ugly and non-compliant (i.e. did not cooperate with their treatment; see Vignette 7.1). And if you are hoping that at least these attitudes would be hidden from their patients, then findings from another study (Brandsma, 2005) are not reassuring: patients' perceptions that their doctors had negative attitudes were even stronger than the negativity reported by these same doctors.

However, in a study of parents' perceptions of attitudes of GPs and other health professionals towards their child's obesity, a range of different responses and attitudes were found (Edmunds, 2005). These were broadly grouped into:

- Helpful: sympathetic response and constructive advice offered, including referrals where appropriate.
- Unaware of how to help.
- Dismissive of the problem: telling patients and their parents not to worry and that the child is likely to grow out of the problem.
- Negative: placing the blame for obesity firmly on the parents and children, and not offering any positive advice.

Despite these attitudes, parents remained positive about their GPs even if they felt that they hadn't been given any useful advice. Although the study only involved interviews with the parents of 40 children, given the difficulty in treating obese children (Activity 6.1) it provided some useful ideas for offering GPs training and support – an issue that will be discussed further in later chapters.

Edmunds's work is not the only UK study to have indicated that many professionals are unclear as to how to deliver effective weight-management advice. And in Australia, although the majority of 752 GPs surveyed considered that they were well prepared to give appropriate advice and treatment to overweight and obese patients, their responses did not entirely bear this out in that they underestimated the use of supportive therapies and the need for active follow-up (Campbell et al., 2000). In common with reports of other studies, the GPs found the management of overweight and obese patients frustrating, many citing patients' low motivation and non-compliance.

As you can see, although doctors recognise the health risks of obesity, they are ambivalent about managing weight loss, and generally believe that obese patients will not make any changes to their lifestyle. These negative attitudes have been shown to cause obese individuals to avoid routine health checks and to delay seeking help for many kinds of health problems. This is especially the case for ailments which may require a physical examination (Fontaine et al., 1998), when patients can feel humiliated, as described in this experience of an emergency admission to hospital (Carryer, 2001)

> They left me there and then poked and prodded around of course and this doctor came in and he said of course you are overweight, and I thought well that was a brilliant deduction, I said 'so it can't possibly be appendicitis then, you know it's just fat?'

There are concerns that negative attitudes from health professionals – e.g. over 30% of nurses saying they would prefer not to treat obese patients (Maroney and Golub, 1992) – result in some obese people receiving poor quality of service.

7.3.2 Mental health

Given the discrimination experienced and the many comorbidities associated with overweight and obesity, it would not be surprising to find that mental health was also affected by the condition (Section 1.4.2). However, in a review of published research, McElroy and colleagues (2004) concluded that most overweight and obese people in the community do not have mood disorders. In fact, depression does not occur more frequently in overweight and obese individuals than it does in those of healthy weight.

◆ What do you understand by the term 'depression'?

◆ 'Depression' may be interpreted differently by different people. You may have defined it as feeling low in mood and possibly not being interested in usual activities. Depending on your previous knowledge and experience, you may think of depression as an illness, or you may think of it as a self-limiting period of feeling unhappy.

It is normal for people to experience a range of negative feelings from time to time and to describe this as being depressed, but **clinical depression** is an illness estimated to affect around 25% of the population at some point in their lives and encompasses a variety of different symptoms:

- psychological: for example, continuous low mood, feelings of hopelessness, low self-esteem, feelings of guilt, lack of motivation and enjoyment, suicidal thoughts
- physical: for example, change in appetite, slowed movement or speech, difficulty sleeping, waking very early, lack of energy
- social: not performing well at work, avoiding social situations, difficulties in home and family life.

Looking at the list above, you can see that one or more of these symptoms might be found in an individual without that person being clinically depressed. **Self-esteem**, our assessment of our attributes and attainments, is often assumed to be low in obese people. In a prospective (cohort) study of children aged 9–10 at recruitment, Strauss (2000) found that, among obese white and Hispanic girls, significantly more had developed low self-esteem by the age of 13–14 when compared with non-obese peers. Those with low self-esteem were also significantly more likely to experience loneliness, sadness and nervousness than obese children with good self-esteem. However, in a survey of studies, Wardle and Cooke (2005) found that: 'effects are moderate at most, and scores rarely fall outside normal ranges' (see Vignette 7.3). Puhl and colleagues (2005) come to similar conclusions about adult obesity and self-esteem.

Vignette 7.3   Charlie is teased

The new sport/biology teacher, Mr Boyce, has introduced an after-school swimming club and Charlie, now 13 years old, is certain of a place on the school team. He swims at least twice a week and attends Mr Boyce's training sessions weekly. He is much fitter and no longer has problems of breathlessness. But he is still tubby. Charlie and his mates are at the leisure centre, playing ball in the pool before the club starts, when his classmate Brad passes by on his way to the café. He yells and waves at the boys, then points at Charlie and cups his hands around his own ribs. Charlie understands (see Figure 7.5), grins and makes a rude gesture. Mr Boyce, who witnessed the whole episode as he came into the leisure centre, notes that Charlie's smile soon goes and that subsequently his swimming is not really up to the usual standard. He resolves to tell Charlie that he is on the team and to discuss a training regime and diet with him.

Many of the studies where self-esteem or depression is reported to be correlated with obesity are based on groups who are seeking treatment for obesity rather than from surveys of the general population. One exception to this (Carpenter et al., 2000) found that the risk of suicidal thoughts and depressive disorders was raised in obese women but in men there was a decreased likelihood of depressive disorders. Other similar findings suggest that the experience of obesity may be differently perceived by men, and particularly in relation to their **body image**.

Body image is a person's perception of their body. If a person has a poor body image, they will see themselves as being unattractive or even repulsive to others, whereas someone with a good body image will see themselves as attractive – or, at least, will feel comfortable with their body. In a review of studies, Schwartz and Brownell (2004) reported that around half of people are in some way dissatisfied with their body. Poor body image is more likely in obese individuals, and the degree of obesity increases the risk of poor body image – that is, the higher a person's BMI, the more likely they are to have a poor body image. Generally speaking, women tend to have poorer body image than men (see Vignette 7.4), and overweight white women are less satisfied with their bodies than are black women of similar weight.

Vignette 7.4   Introducing Linda

Linda is 51 and has been married to Doug for 23 years. They have two children, aged 17 and 15, both of whom live at home. Linda is 1.63 m tall and weighs 110 kg, and has been overweight for as long as she can remember. Both her parents were alcoholics and would give her food to keep her quiet and out of their way. She remembers many occasions when her father would hit her when he was drunk. She has very low self-esteem and thinks she is worthless because she has not worked since she had the children. For the past 4 years, she has seldom left the house. She knows the children prefer her not to be seen at school functions. Doug and the children try to cheer her up by bringing her foods she likes: she is particularly fond of crisps and will eat up to ten packets in one go.

Linda avoids looking in the mirror but she knows she needs to lose weight. She has tried countless diets but becomes disheartened very quickly and always gives up after a few weeks. She is embarrassed to go to the gym or swimming pool, and anyway she has arthritis in her back, hips and knees, which makes it difficult for her to exercise.

Linda eventually plucks up the courage to see her doctor about her weight because the family is flying to Florida on holiday in a few months' time but she is worried she won't fit into an aeroplane seat. She asks her doctor for medication to help her lose weight but her doctor refuses to give her anything, saying that there is no magic cure; she simply needs to eat less and exercise more.

You will read more about Linda later.

◆ How do you think Linda perceives herself and her body?

◆ Linda is not happy with her body. She avoids looking at herself in the mirror and has very low self-esteem. She feels too embarrassed to go to the gym or go swimming, and is worried about whether or not she will fit into an aeroplane seat.

◆ Now re-read Vignette 7.1. Does Angus have a poor body image?

◆ It would appear that Angus does not have a poor body image. He has several friends who are a similar size to him and he is not concerned about his weight. For him, this is normal. He accepts his body as it is and does not particularly see the need to change.

A study of over 5000 people in the USA (Paeratakul et al., 2002) found that men were less likely to be dissatisfied with their bodies, concerned about body weight, or to wish to lose weight.

◆ Look at Figure 7.7a. What do you notice about men's perception of their size?

◆ There are considerable numbers of overweight and obese men who do not see themselves as being overweight.

This finding is repeated in other studies (Figure 7.8) and raises concerns about men's health at a time when the numbers who are overweight and obese continues to increase.

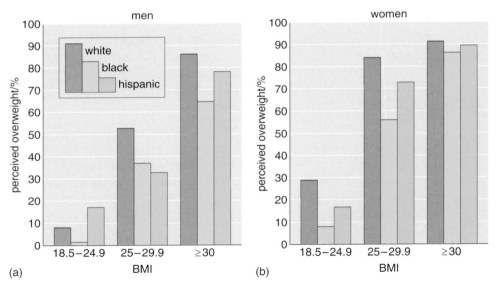

Figure 7.7 Perceived weight status among overweight and obese (a) men and (b) women.

7.3.3 Eating disorders

An eating disorder is a compulsion to eat, or to avoid eating, in a way which disturbs physical and mental health. So it might seem that obesity could be a consequence of an eating disorder but not vice versa. However, as with the arguments that cause and effect are unclear where depression and obesity are associated (Section 1.4.2), there are similar arguments relating to **binge eating disorder (BED)** and obesity. Binge eaters typically rapidly consume much more food than most people could manage, eating until they feel uncomfortably full.

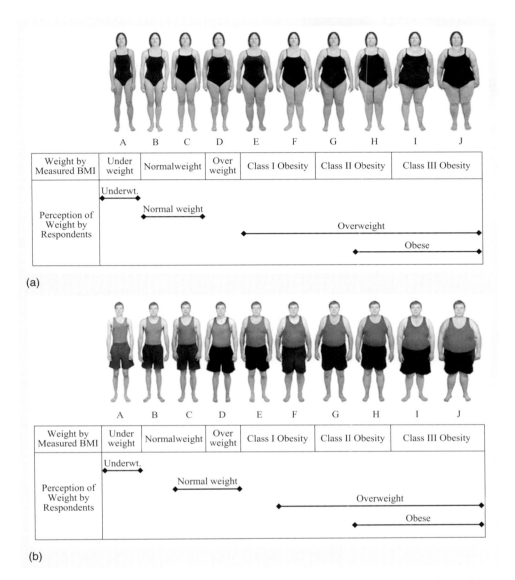

Figure 7.8 Body size guide images for (a) women and (b) men, together with the associated weight classification, and the perceptions of size as judged by the majority of respondents.

They eat in spite of not feeling hungry and feel unable to control themselves when eating. After eating, binge eaters typically feel guilty or disgusted with themselves.

BED is more common in obese individuals than in non-obese ones and is often particularly high in those who repeatedly attempt to lose weight. A number of studies associate BED with weight cycling (see Box 7.1). A prospective study of 2516 adolescents found that having been teased about their weight was a good predictor of BED at a follow-up 5 years later (Haines et al., 2006). So it is at least a possibility that BED may develop in some people as a consequence of the way they cope with low self-esteem associated with their obesity.

Box 7.1 Weight cycling

Weight cycling describes the repeated cycles of losing and gaining weight as a consequence of repeated dieting and associated weight rebound afterwards. Several studies have raised concerns that this behaviour has a detrimental effect, increasing the likelihood of the development of overweight in the longer term and hence contributing towards the current obesity epidemic.

For example, one study (Field et al., 2004) used a very large database of female nurses (The Nurses Health Study II) to compare the weight gained by individuals who gained weight in different ways with individuals whose weight remained stable over the 4-year period of the study (Table 7.2). So the groups were:

- stable weight (non-cycling)
- weight gain (non-cycling)
- mild weight cycling: those who reported intentionally losing 4.5 kg or more, 3 or more times over the course of the study

- severe weight cycling: those who reported intentionally losing 9 kg or more, 3 or more times over the course of the study.

Table 7.2 Weight gain over 4 years following episodes of weight stability, weight gain, mild weight cycling and severe weight cycling.

Group	Weight gain/kg
stable weight (non-cycling)	0.0
weight gain (non-cycling)	1.5
mild weight cycling	4.5
severe weight cycling	6.2

Most of the nurses had used diet to try to control their weight, but the severe weight cyclers were significantly more likely than the other groups to have used unhealthy diets and not to have tried using exercise to support their efforts. (Different types of diet and exercise will be considered in Chapter 8.)

7.3.4 Social and economic consequences of obesity

Various surveys have found that overweight and obese people may feel so unattractive that they avoid social situations. Their feelings may not be as extreme as those shown in Table 7.3, which were the responses of a group of 57 morbidly obese individuals who were about to undergo weight loss surgery, but they can contribute to unsatisfactory social and workplace relationships. Adolescents – particularly boys – are less likely to want an obese partner. At college/university, obese women are deemed less sexually attractive and are romantically dated less often. Obese women are more likely to marry later and to move to a lower socioeconomic group over their lifetime. With less good educational and employment prospects, it is unsurprising that obesity is associated with a lower socioeconomic status (Section 1.6.4).

◆ What personal economic costs can you think of that might be a consequence of obesity, regardless of social status?

◆ There are many different answers which you could give here:

- the cost of health care: even countries with state-funded health care often charge for medication prescriptions

- life insurance may also be more expensive and difficult to obtain (see Vignette 7.1)

Table 7.3 Responses of morbidly obese patients about to undergo weight loss surgery to questions concerning their weight and psychosocial functioning.

	Response/%			
	Very unattractive	Somewhat unattractive	Somewhat attractive	Very attractive
How physically attractive do you feel, taking everything into account?	96.5	3.5	0	0
	Always	**Usually**	**Sometimes**	**Never**
At work, people talk behind my back and have a negative attitude towards me because of my weight	80.7	10.5	3.5	5.3
I feel that my weight has negatively affected whether or not I have been hired for a job	67.3	20.4	10.2	2.2
I do not like to be seen in public	66.7	17.5	14.0	1.8
I feel that I have been treated disrespectfully by the medical profession because of my weight	45.5	32.7	16.4	5.5

- clothes may be more expensive (and the choice limited)
- the cost of food may be high (see Vignette 2.1)
- because of poor mobility, it may be necessary to pay others to help with domestic jobs
- lower earning potential and/or inadequate pension: maybe having to live on benefits.

◆ Re-read Vignette 7.4. Does Linda face any social or economic problems as a result of being obese?

◆ Linda is very isolated. She does not have a job and she feels embarrassed to leave the house, even to attend school meetings. She is also concerned about whether or not she will fit into a normal-sized seat on an aeroplane. She is thus worried about the way in which she is perceived by other people and her ability to cope in an environment catering mainly for non-obese individuals.

Just as obese individuals have arrived at obesity via different routes and have different – or no – comorbidities, so they exhibit a variety of attitudes and feelings towards their own weight as well as a variety of coping strategies.

◆ How would you describe the attitudes of Angus, Nazneen and Linda towards their weight?

◆ Angus *denies* that weight is a problem – he has many friends of a similar build and was surprised to learn about the possible effects his weight is having on his health. Linda is deeply unhappy about her weight, but this seems to be for *cosmetic* reasons rather than medical ones, and she is prompted to seek medical help only when faced with a difficult social situation. Nazneen recognises that her weight is a *health* problem – she understands the impact it could have on herself and her baby.

Denial

Denial is the refusal to acknowledge a problem or a responsibility. Section 1.1 stated that there are obese people who deny that obesity is a disease. There are 'fat acceptance' groups in Europe and the USA. Angus denies that he has a problem – he is happy with his weight and has no particular motivation to lose weight. As an ex-rugby player, he may well see himself as big and strong. Indeed, some of the overweight individuals in Figure 7.8 (as judged by BMI) who do not see themselves as overweight may be muscular and very active. As you will read in Chapter 8, to be fat and fit is less unhealthy than to be slim and sedentary. However, many overweight people do not maintain fitness in later life and subsequently experience comorbidities.

Although certain individuals are more likely to become obese than others (see Chapter 6), some obese individuals deny any responsibility for their weight. Others claim to live a healthy lifestyle, whereas studies assessing intake of food have frequently found that such individuals under-report the amounts of food they eat (Section 1.4.1).

A cosmetic issue

You have already learned how obesity increases the likelihood of a poor body image – for many obese individuals, this is the key motivation behind any desire to lose weight. More women request weight-loss surgery for cosmetic reasons than for health reasons; hardly surprising when you have only to read any women's magazine to see how celebrities are frequently criticised and ridiculed for putting on weight and applauded for losing it.

As you read in Vignette 7.4, Linda's main reason for seeking medical help in losing weight is because she is due to fly to Florida and is concerned she may not fit into an aeroplane seat. Many social settings are designed for people with a smaller body size – this includes trips to the cinema or theatre, using public transport and dining in a restaurant. It can be a humiliating experience for an individual to find they cannot fit into a seat or cannot get up from a seat (Figure 7.9a); in the USA, there have been several law cases where obese individuals have sued companies for causing emotional distress after being asked to pay for two seats (Puhl and Brownell, 2001).

Figure 7.9 (a) Seats are too small for some people. (b) An example of a wheelchair and an ambulance for larger people, with standard-sized versions for scale.

(a)

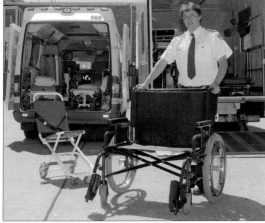

(b)

Health

Of the three individuals in the vignettes featured in this chapter, only Nazneen seems to recognise the health risks of her weight. Public health interventions targeting obesity have been initiated because of this risk to health. There is a shortage of specialist health care; for example, there are only nine specialist obesity clinics in the UK, each with a waiting list that currently (2008) must not exceed 5 weeks, so many potential patients have to be turned away. From reading this chapter, you will be aware that there are difficulties and barriers to implementing any strategy and that there will be no 'one size fits all' type of solutions for such a heterogeneous group of people with different levels of awareness and reactions to the stigmatising environment.

It is important that the negative views and associations of obesity held by both the general public and some health professionals are challenged. It is equally important not to ignore or accept the problem – the risks of obesity need to be acknowledged, together with the need for weight loss, but without the associated negative stigma which currently exists towards obesity. The next two chapters look at some practical solutions at the level of the individual, then Chapter 10 looks at how society, as a whole, might respond to the challenge of obesity.

7.4 Summary of Chapter 7

7.1 Overweight and obesity increase the risk of different medical conditions as a result of various changes in the body.

7.2 Even conditions such as sleep apnoea or osteoarthritis, where the presence of fat crowding other organs or adding to the weight of the body might seem to be only a physical problem, result in physiological changes to the body.

7.3 Failure to maintain insulin homeostasis underlies many of the medical consequences of obesity.

7.4 Obese people face discrimination and negative attitudes – even from health care professionals.

7.5 Despite the stigma of obesity, research shows that most overweight and obese individuals do not have mood disorders such as depression.

7.6 Depression and low self-esteem are found more often in those who are morbidly obese and seeking help.

7.7 Obese and overweight people hold a variety of different attitudes to their weight which affects their feelings about themselves.

Learning outcomes for Chapter 7

LO 7.1 Define and use, or recognise definitions and applications of, each of the terms printed in **bold** in the text.

LO 7.2 Outline the potential medical consequences of obesity and understand the reasons why overweight and obese individuals are at increased risk of these conditions.

LO 7.3   Discuss the psychological consequences of obesity and understand the potential impact of obesity in social settings, such as education and employment.

LO 7.4   Describe the different attitudes held by health care professionals towards obese individuals and the effect this may have on the treatment of obesity.

LO 7.5   Recognise a range of different attitudes towards obesity held by obese individuals.

Self-assessment questions for Chapter 7

Question 7.1 (LO 7.2)

Explain why an obese individual is at greater risk of experiencing periods of breathlessness than a non-obese individual.

Question 7.2 (LO 7.3)

Do you think Linda has BED? Give reasons for your answer.

Question 7.3 (LO 7.4)

How might the attitude of a doctor or nurse affect the desire of an overweight individual to lose weight? Use Angus, Nazneen and Linda to illustrate your answer.

Question 7.4 (LO 7.5)

Looking at Figure 7.7, which category do you think would have described Nazneen prior to her discovery that she had impaired glucose tolerance? In relation to conceiving and maintaining her pregnancy, what reference to her weight has been made by the medical profession?

CHAPTER 8

REDUCING OBESITY: DIET, EXERCISE AND LIFESTYLE

I've been on a constant diet for the last two decades. I've lost a total of 789 pounds. By all accounts, I should be hanging from a charm bracelet.

Erma Bombeck, US humourist, 1927–1996

8.1   Introduction

Earlier chapters in this book have emphasised that an increase in body weight can only occur when there is a positive energy balance, when energy intake exceeds energy output. From this simple fact, it follows that there are two obvious and complementary approaches to reducing weight: reducing food intake or increasing physical activity. Indeed, it is possible to calculate the likely consequences for weight that will result from changes in food intake or physical activity. However, as you may be aware from your own experience or that of family or friends, many people try reducing food intake by dieting or try increasing physical activity through exercise, but are no lighter a year or two later. No doubt that was also the experience of Erma Bombeck, quoted at the beginning of this chapter. Perhaps they all failed because some diets are better than others and they chose the wrong one? Or perhaps exercise and the right diet need to be combined? Later in this chapter, you will look at some of the scientific studies that have addressed these and similar questions. You will also look at some of the ways in which individuals might be enabled to change their behaviour in the longer term to achieve weight loss.

The chapter begins with a few calculations to illustrate what should happen following changes to energy intake or energy expenditure.

8.2   Energy balance revisited

The components of the energy equation were introduced in Chapter 2.

◇   In terms of these components, how can a negative energy balance be achieved?

◆   A negative energy balance can be achieved when energy intake is less than energy expended.

Consider the effects of a negative energy balance on a hypothetical overweight person, Paul. He is just under 1.83 m (6 ft) tall and weighs 90 kg (14 stone 2 lb). He has weighed 90 kg for 4 years, which means his current energy intake exactly balances his energy output. He decides to take regular exercise. He replaces 60 minutes of the day when he would be sitting with some brisk walking. He does 40 minutes walking at lunch time and 20 minutes walking in the evening and walks at 3 mph (miles per hour). We can calculate the effect that this additional energy expenditure will have on his weight. But the calculation requires one further piece of information in addition to Paul's weight and the duration of the activity: the energy expenditure of the activity. The easiest value to use is the **metabolic equivalent (MET)**, partly because MET values (see Table 8.1; Ainsworth, 2002) have been established for countless activities and partly because the MET value is simple to use. Sitting quietly is rated at 1 MET. Thus this is the measure used for BMR (basal metabolic rate); recall from Section 2.7 that BMR values are remarkably constant when related to lean body mass, although they may reduce after substantial loss of body weight while dieting.

1 MET is equivalent to a metabolic rate of 1 kilocalorie per kilogram of body weight per hour.

Table 8.1 The energy cost of some physical activities expressed as METs.

Activity	MET
Walking at 3 mph on a level and firm surface	3.3
Walking at 3.5 mph on a level and firm surface	3.8
Walking at 3.5 mph uphill	6.0
Jogging	7.0
Running at 6 mph	10.0
Tennis, singles	8.0
Cycling for leisure at 12 mph, moderate effort	8.0
Soccer, friendly game	7.0

Table 8.1 lists a range of MET values for various activities. These values are for TEE (total energy expenditure; Section 2.1). So if you wanted to know what Paul's additional energy expenditure was for an hour of jogging, you would take the value of 7.0 METs from Table 8.1 and subtract from it the 1.0 MET that represents his BMR. Thus during the hour of jogging, Paul's additional rate of energy expenditure, over and above his BMR that would have happened anyway, is:

$$7.0 \text{ kcal kg}^{-1} \text{ h}^{-1} - 1.0 \text{ kcal kg}^{-1} \text{ h}^{-1} = 6.0 \text{ kcal kg}^{-1} \text{ h}^{-1}$$

◆ What is Paul's additional energy expenditure for an hour walking on a firm level surface, over and above his BMR?

◆ $3.3 \text{ kcal kg}^{-1} \text{ h}^{-1} - 1.0 \text{ kcal kg}^{-1} \text{ h}^{-1} = 2.3 \text{ kcal kg}^{-1} \text{ h}^{-1}$

So during this activity, Paul expends an additional $2.3 \text{ kcal kg}^{-1} \text{ h}^{-1}$.

◆ How many additional kilocalories in total will Paul use each day walking for 60 minutes instead of sitting at rest?

◆ $2.3 \text{ kcal kg}^{-1} \text{ h}^{-1} \times 90 \text{ kg} \times 1 \text{ h d}^{-1} = 207 \text{ kcal d}^{-1}$

◆ If Paul does this every day for one year (365 days), how much additional energy will he expend over the year?

◆ He will expend an additional

365 d × 207 kcal d^{-1} = 75 555 kcal

Assuming that Paul is not changing his energy intake, this additional energy must come from fat stored in adipose tissue. Each kilogram of adipose tissue can release about 7700 kcal of energy.

◆ Why do you suppose this energy value for adipose tissue, equivalent to 7.7 kcal g^{-1}, is less than the 9 kcal g^{-1} value given in Table 2.1 for fat?

◆ Adipose tissue comprises more than just fat. For example, each adipocyte contains a nucleus and mitochondria in the cytosol; then there is the material that holds adipocytes together, together with the blood supply and water.

Hence the fat content of adipose tissue is about 85% of the mass of adipose tissue. The value of 7.7 kcal g^{-1} is an approximation, but a reasonable one for our purposes as it is 85% of 9 kcal g^{-1}.

Returning to Paul; if 1 kg of his adipose tissue can release 7700 kcal then, in order to generate the 75 555 kcal needed for the physical exercise over a full year, the amount of adipose tissue Paul must have used is:

$$\frac{75\,555 \text{ kcal}}{7700 \text{ kcal per kg}} = 9.8 \text{ kg}$$

Therefore, by walking for 1 hour a day, Paul would lose about 9.8 kg in a year.

A similar type of calculation can be done for a reduction in energy intake. Suppose Paul decided to remove the packet of crisps from his lunch box each weekday. What would be the effect on his weight over the course of a year?

A proprietary brand of crisps details their single-serving, 34.5 g bag of salt and vinegar crisps as containing 181 kcal. Hence, per year, that would be:

181 kcal × 5 × 52 weeks = 47 060 kcal y^{-1}

◆ How much would eating 47 060 kcal over the year add to Paul's weight?

◆ The amount of adipose tissue added to Paul's body would be:

$$\frac{47\,060 \text{ kcal}}{7700 \text{ kcal kg}^{-1}} = 6.1 \text{ kg}$$

In some nutritional information, energy values are quoted in kJ – these can be converted to kcal by dividing by 4.2.

However, Paul is now no longer eating this daily packet of crisps, so because he was in energy balance before he started on this new regime, he will lose an additional 6.1 kg of weight in a year. The combined effect of these two changes – walking an hour a day and no crisps at lunch – would be an impressive 9.8 + 6.1 = 15.9 kg weight loss over the course of a year.

The two examples are mathematically correct, but they do depend on some approximations and a rather important assumption.

◆ What is the important assumption? (Hint: What would happen if you removed an item from your lunch box each day?)

◆ The important assumption in the figures is that nothing else changes. If you removed an item from your lunch box, you might change something else: put more filling in your sandwich or add an extra piece of fruit, for instance.

Having read Chapter 6, you may also have noticed the assumption that the same values apply to different people, an assumption explored in Box 8.1.

Despite these assumptions, the data do illustrate that weight change should be possible by creating a negative energy balance, through increasing physical activity levels and/or reducing the calorific value of your meals. However, it has often been noted that the degree of weight loss during dieting may be less than expected from these types of calculations. There are likely to be two reasons for this. It is known that the reduction of body mass during dieting is associated with a reduction in BMR (Section 2.7). In addition, people who are dieting may become less active than before, again reducing the effect of dieting on their overall energy balance.

Box 8.1 Approximations used in energy calculations

The number of kcal used in any particular activity varies from person to person, depending on their actual BMR, their weight and the effort put into the activity. There are many energy expenditure tables available on the internet. They provide values for diverse activities such as walking, ironing and swimming. They usually provide different values for the intensity of the activity (e.g. walking speed), and the weight of the participant. (Note that many diet sites, particularly those based in the USA, refer to calories when they mean kilocalories.) Although there is general consistency across these sites, the actual values vary, depending on exactly how the data were obtained. Some people think that MET values underestimate energy expenditure, especially that of larger people. Also, some calculations, like those above, may involve the removal of BMR. Such calculations would be appropriate where the participant wants to know how much *extra* energy they have burnt off: if they had been sitting still for an hour, they would have burnt off an hour's worth of energy at their BMR, so the extra is what they actually burnt off doing the exercise minus the hour's worth at their BMR. Remember that using these MET values can give only an approximation for any one individual.

8.3   Reducing energy intake: 'going on a diet'

For most people, the initial response to the question of how to lose weight is to 'go on a diet'. As you know from Chapter 2, energy is derived from your diet, but your diet is actually *everything* you eat and drink. If someone says they are 'on a diet', they usually mean that they are following a plan whereby they alter their normal eating habits to achieve weight loss. This section reviews different kinds of diet for weight loss, from the most extreme to the more frequently followed. It is important to note that medical supervision is a very important aspect of the more severe diets that are described here. This section also provides a brief overview of scientific studies that have examined the effectiveness of different forms of dieting in both the short and the long term.

8.3.1   Starvation

The most drastic form of diet for weight loss would be simply to stop eating and to drink only water. Although this will promote weight loss, it is an undesirable strategy and is not endorsed by medical professionals. After a few days, there will have been a considerable loss of fluid – and some protein loss too, as muscles start to waste. Beyond 14 days, there will be continued loss of lean tissue as well as fat. The risk of sudden death is greatly raised in individuals who completely abstain from food for a prolonged period and is especially high in obese individuals who are not eating. The most important reason for this is that many minerals are slowly but steadily lost from the body through the urine. During starvation, these minerals are not replaced, leading to a shortage. Two such minerals, magnesium and potassium, are important for a healthy heart, and their loss contributes to disturbed heart function.

◆ In obesity, there is a structural change to the heart; what is this change and why does it occur?

◆ The size of the left ventricle which pumps blood to the major body organs is increased because body mass is greatly increased, as is the volume of blood required to oxygenate the increased amount of tissue. The left ventricle therefore has to work harder in order to pump the increased amount of blood around the increased amount of tissue, and it enlarges as a consequence (see Section 7.2.2).

The increase in size of the left ventricle disturbs normal heart beats (contractions) and increases the likelihood of abnormal and rapid contractions, known as ventricular fibrillation. The levels of minerals that are dissolved in fluids in and around the cardiac muscle tissue have to be maintained around optimum values to maintain the pattern of heart contraction. Decreases in mineral levels also increase the likelihood of ventricular fibrillation.

8.3.2   Very-low-calorie diets

Very-low-calorie diets (VLCDs) are used under medical supervision as a way of inducing rapid weight loss while reducing the risk of sudden death associated with starvation. VLCDs are usually defined as ones that provide between 400 and 800 kcal d^{-1}. Typical energy requirements to maintain constant body weight are affected by a variety of factors, as described in Chapter 6. For instance, young, active individuals will have a higher daily energy requirement than older, more sedentary individuals. Also, men generally have a higher proportion of lean tissue than women. Because lean tissue, such as muscle, has a higher metabolic rate than adipose tissue, the overall daily energy requirement for men tends to be greater than for women. A VLCD of 800 kcal d⁻¹ will provide half the usual energy requirement for a woman who takes little exercise and requires about 1600 kcal d⁻¹, but will provide a much smaller fraction for a more active man whose requirement is greater than 2500 kcal d⁻¹. As a consequence, it is sometimes argued that a VLCD should instead be defined as one that provides an individual with less than half their average daily energy requirement.

VLCDs should only be used under medical supervision and are most appropriate for very obese individuals who need to lose weight rapidly because of considerable, imminent health risks. VLCDs may also be used by individuals who are preparing for the types of surgery described in Chapter 9. The most common technique is to use formulated liquid diets to provide complete nutrition over a 10- to 12-week period. During this time, patients would expect to lose 1.5–2.5 kg per week. They will then begin to reintroduce normal foods, with an emphasis on a permanent change in eating habits towards foods lower in fat and a diet containing a higher proportion of fruit and vegetables. A particular risk during this period is the development of binge eating (see Section 7.3.3), which can potentially become established as a more permanent eating disorder. Although feelings of intense hunger usually disappear quickly after initiating a VLCD, the patient may be left feeling tired and unable to concentrate. These issues may be particularly severe if the energy content of the diet is at the lower end of the VLCD range. Indeed, the evidence suggests that lowering the energy content much below 800 kcal d⁻¹ produces little enhancement of weight loss, both because of a lowering of BMR and because individuals become very inactive, reducing their energy output. Clinical depression is a further risk, so VLCDs are not recommended for those with mental health problems. There are a range of other side effects associated with VLCDs; for example, there is an increased risk of developing gallstones (see Section 7.2.5).

In summary, in the short term, VLCDs, under medical supervision, are an effective way of inducing rapid weight loss over a relatively short period of 2–4 months, but are associated with a wide range of undesirable side effects.

8.3.3   Low-calorie diets

Low-calorie diets are conventionally defined as ones in which energy intake has been reduced by some 500–1000 kcal d⁻¹ below that required to maintain a stable weight.

◆ What weight loss would you expect over a 4-week period in which energy intake is 1000 kcal d⁻¹ lower than required for a stable weight? (Hint: Look back at Section 8.2 for the data that are needed here.)

◆ A 4-week period is 28 days long. The energy intake is 1000 kcal d⁻¹ lower than required, so the energy deficit would be 28 000 kcal over the 4 weeks. Each deficit of 7700 kcal might be expected to lead to the loss of about 1 kg of body weight, so the total weight loss should be:

$$\frac{28\,000 \text{ kcal}}{7700 \text{ kcal kg}^{-1}} = 3.6 \text{ kg}$$

In practice, weight loss of between 0.5 and 1 kg per week results from this kind of diet, which is consistent with the calculated estimate.

It is much easier to maintain a balanced and healthy diet with the relatively smaller reductions in total energy intake used in low-calorie diets compared with in VLCDs.

◆ What is meant by a balanced diet?

◆ A balanced diet is one that provides the six key nutrient groups (the macronutrients and micronutrients) in appropriate amounts for health (see Section 2.3). The WHO advises that a healthy balanced intake of the energy-providing macronutrients is: 55–75% carbohydrates; 10–15% proteins and no more than 10% fat.

◆ Why is it also important to maintain an appropriate level of micronutrient intake?

◆ Many micronutrients – including minerals such as potassium and magnesium – cannot be stored in the body, so they must be continuously available from the diet.

One approach to maintaining a low-calorie but balanced diet is to replace one meal with a commercial liquid meal replacement using products such as Slim-Fast®, Complan® or Ensure® (Figure 8.1). There have been a number of experiments that compare individuals dieting with or without liquid meal replacement of one meal per day. The consensus from such studies is that, on average, the meal replacement group lose about 2.5 kg more after 3 months.

Figure 8.1 A selection of liquid meal-replacement slimming products.

◆ Suggest some possible explanations for the greater effectiveness of diets involving liquid meal replacement.

◆ A number of possible reasons are shown below; you may have thought of others.

- The liquid meal replacement consists of a single flavour. Even if that flavour is attractive, there is no variety within the meal to enhance appetite and increase intake (see Section 5.4.1).

- Liquid meal replacements are low in calories for their volume. The greater degree of stomach distension for a given number of calories might suppress appetite (see Section 5.3.1).

- Liquid meal replacements are expensive – relative to other pre-prepared or canned drinks, at least. Having to purchase the diet may encourage individuals to commit to their broader diet plan.

- Liquid meal replacement reduces the temptation to snack on inappropriate foods.

8.3.4 Popular diets

Many of us have looked at popular diet books. Indeed, according to statistics on the borrowing of books from UK public libraries in 2006/7, 4 of the 10 most borrowed non-fiction titles were diet books. At this point, we (the authors) considered several strategies. We could provide you with an (almost endless) review of different diets and their supposed justification. Perhaps we might instead present the OU Diet Plan, hoping to emulate the success of these other books! But we rejected these possibilities and instead will look at four diets that were the subject of a large-scale and well-controlled comparative trial (Gardner et al., 2007): the Atkins, Zone, Ornish and LEARN diets. The key difference between these diets is their macronutrient balance, and this distinguishes them from the calorie-controlled diets considered in previous sections.

The Atkins diet. This diet, devised in the 1960s, became really popular in the early 2000s. It advocates the consumption of a high-protein, low-carbohydrate diet. Most individuals find that the high-protein element of the diet leads them to reduce their overall energy intake. Surprisingly, it is not that enjoyable to maintain this diet, and it is rather expensive. Robert Atkins, who was very overweight himself, devised the diet shortly after leaving medical school, based on his supposition that the high content of refined carbohydrates in modern diets (e.g. products such as white bread and white sugar) is a major cause of obesity. By greatly restricting the consumption of carbohydrate, Atkins suggested that the body would burn fat inefficiently, metabolising it to ketones as it ran out of easily available glucose.

◆ Why are fatty acids metabolised to ketones when glucose is in short supply? (Hint: Refer back to Box 4.1.)

◆ The initial stage of fatty acid metabolism yields acetyl CoA, which can enter the TCA cycle (see Figure 2.6). But when glucose is in short supply, the TCA cycle cannot function with sufficient efficiency to use all the acetyl CoA, and the surplus is metabolised to ketones.

See Section 5.4.1 for a reminder of what is meant by a 'well controlled trial'.

Although ketones are used as a fuel source, with energy being available from the breaking down of chemical bonds as they are metabolised, they are incompletely metabolised and the molecules that are excreted still have energy within them that has not been released. Atkins maintained that this inefficient fat metabolism would lead to a greater weight loss than would be expected from the overall reduction in energy intake typical of this diet.

The Zone diet. The low-carbohydrate Zone diet was made popular in the mid-1990s by the paediatrician (children's doctor) Barry Sears. This diet advocates a 30% : 40% : 30% distribution of protein, carbohydrate and fat in the diet. Proponents of this approach suggest that this ratio is beneficial in terms of maintaining an appropriate balance between insulin and glucagon release. The theory is that by keeping these hormones at the right levels (i.e. in the zone), the body regulates fat more efficiently.

◆ What are the effects of these two hormones on fat metabolism?

◆ Insulin encourages the storage of fat by stimulating the synthesis of TAGs from glycerol and fatty acids and inhibiting their breakdown (see Figure 4.9). By contrast, glucagon stimulates the release of fatty acids and glycerol from the TAGs stored in adipose tissue (see Figure 4.11).

The Ornish diet. This diet, promoted by Dean Ornish to reverse heart disease, is almost the inverse of the Atkins diet. It is a vegetarian diet in which high-protein foods such as meat, fish and nuts are virtually prohibited, whereas fruit and grains are strongly encouraged. The distribution of protein, carbohydrate and fat on this diet is approximately 20% : 70% : 10%. The rationale is that heart disease is associated with the consumption of animal protein and saturated fats. It is not primarily a weight-loss diet, but many people following it found that they did lose weight.

The LEARN diet. LEARN stands for lifestyle, exercise, attitudes, relationships and nutrition. The diet was developed as one component of the Lifestyle program by the LEARN institute and, like the Ornish diet, advocates reduced fat (not more than 10%) with high carbohydrate (60%) and moderate protein (30%).

The four diets described above were compared by following the progress of 311 women over a year. The study recruited women with BMIs of between 27 and 40 and randomly allocated each to one of four groups. Each group undertook an 8-week initial treatment period with weekly meetings during which they worked through a book explaining the principles of their particular diet plan. They were all also taught a variety of behavioural techniques, again taken from the book describing their plan, that might help them persist with their diet. They were then followed up over a further 10 months. The data for weight loss from this trial are plotted in Figure 8.2.

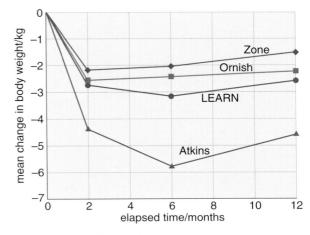

Figure 8.2 Weight loss recorded over a year from participants following four different weight loss diets.

◈ What features do the weight-loss data for the different diets have in common?

◆ Weight loss is most rapid in the first few weeks and months, after which weight loss decreases, stops entirely and may even reverse.

◈ What feature distinguishes the weight-loss data for the four diets?

◆ Three of the diets seem to result in similar patterns of weight loss, but it appears that the Atkins diet was more effective in producing and maintaining loss of weight within the time period of this study.

This last result is rather surprising. There have been a number of similar experiments in which the participants have been randomly allocated to either low-fat, high-carbohydrate diets (such as the Ornish or LEARN diets) or to those that emphasise low carbohydrate (including the Atkins and Zone diets). The broad conclusion from these studies has been that there is little to choose between these two approaches in terms of weight loss.

◈ What possible reasons might explain the greater weight loss with the Atkins diet in this particular study?

◆ There are several explanations given below; you might have thought of others.

- Perhaps the participants in the Atkins group actually ate less, in terms of total energy intake, than those in the other groups.

- Perhaps the women in the Atkins group were more self-disciplined or more responsive to the behavioural techniques.

- Perhaps metabolising fat to ketones really does promote greater weight loss, at least in the overweight women who were recruited onto this study.

The authors of this study estimated the energy intake of the women in the different diet conditions by asking them to recall what they had eaten over specific 24-hour periods. Although energy intake did decrease after they were enrolled onto the different diet programmes, there was no evidence to suggest a greater reduction in energy intake in the Atkins diet group. However, it is hard to recall accurately what has been eaten and the precise quantity over a period of 24 hours, as Activity 8.1 may demonstrate.

Activity 8.1   Food diaries

Allow 10 minutes for this activity on each of two days

Try to recall and write down exactly what and how much you have eaten over the last 24 hours. When you are reasonably satisfied that your list is complete, begin a record of what you eat over the next 24 hours. Once you have completed this, compare the two lists. Which method do you think is the most reliable?

Now look at the comments on this activity at the end of this book.

As has already been mentioned, the differences between diets found by Gardner have not been replicated in other studies. So, at least for the moment, it is unclear whether certain types of diets do have advantages over others. What is certain is that all these diets do lead to weight loss in the short term. There were also considerable differences *between* individuals on the same diet in terms of weight lost, and many clinicians believe that the best approach is to negotiate a diet with each individual – choosing one to which the patient can commit, and which will lead to a permanent change in behaviour. Indeed, subsequent additional data from the Gardner study showed that, regardless of diet type, better compliance with the diet was associated with greater weight loss.

◆ What tends to happen to dieters over longer periods of time?

◆ There is a tendency to regain the weight that had been lost.

The pattern of weight gain for a whole group is likely to mask large variations between individuals. Some participants will maintain their weight loss and others will regain the lost weight – and maybe even exceed their initial weight. The phenomenon of *weight cycling* was noted in Box 7.1 and is possibly a consequence of the mindset that associates 'going on a diet' with 'going on a journey'. In both cases, there is an end point: the arrival at the desired destination (or weight), after which behaviour can change, reverting to previous habits. More will be said about this in Section 8.6.

Before moving on from these diets that manipulate macronutrient levels, take a look at Table 8.3.

◆ Do any of these macronutrient diets match the typical diet of a Western community shown in Figure 2.2?

◆ No, the typical Western diet contains more fat than these diets, with the exception of the Zone diet, which advocates more fat and more protein but less carbohydrate.

Table 8.3 The balance of macronutrients: comparison of WHO guidelines with some popular diet plans.

Guidelines/diet plan	Composition of macronutrients/%		
	carbohydrate	protein	fat
WHO	55–75	10–15	10
Atkins	low	high	high
Zone	40	30	30
Ornish	70	20	10
LEARN	60	30	10

◆ Which of the four diets is the best in terms of promoting overall health?

◆ The Ornish diet comes closest to the WHO guidelines for a healthy, balanced diet.

The other three diets were devised primarily as weight-loss diets. Weight-loss diets often suggest that micronutrient supplements should be taken and that the balance of the diet be altered once the desired weight is reached. It may be that some people abandon diets because they feel that their target weight will never be reached and that, in the meantime, they are neither enjoying what they are eating nor are they feeling in good health. Nevertheless, this section has shown that diets do work – at least in the short term.

8.3.5   Dieting and long-term weight loss

It is clear that dieting can produce significant weight loss over a few weeks or months, although there may be weight regain in subsequent months. Weight loss of 3–5% is sufficient to produce worthwhile gains in health terms. Control of blood glucose levels and insulin sensitivity are improved, which will lead to a reduction in the risk of diabetes. Similar improvements are likely in relation to the other health risks of obesity. But what is the likelihood that weight lost while dieting can be maintained over the ensuing years?

Before we consider the results from studies that have addressed this issue, let us first consider what kind of evidence is required to come to a reliable conclusion about the effectiveness of dieting in the longer term. There is general agreement that good studies need to:

- have a large number of participants. This is also called having a large *sample size* and means that the conclusions they come to are not likely to be distorted by very unusual data from a few individuals.

- have been able to follow the sample for a prolonged period of time. Here, a minimum of 1 year is essential, but studies that extend to 2–5 years are far preferable.

- retain the great majority of participants over the entire study period. If a substantial proportion of participants is lost during the study, there is the possibility that those who remain do not provide a good representation of what has happened to the whole sample. For example, it may be that the participants for whom the diet was not working are the ones who give up and drop out of (are lost to) the study.

The process of collecting together a number of studies addressing the same issue and then combining them in such a way as to get a more accurate estimate of the effectiveness of a treatment is known as a *meta-analysis*. One analysis of this type (Tsai and Wadden, 2006) summarised six studies in which VLCDs and low-calorie diets were compared. In the short term, there was no doubt about the greater effectiveness of the VLCDs. At about 12 weeks (the individual studies varied in the timings between measurements), the VLCD groups had lost an average of 16.1% of their initial body weight, whereas the low-calorie diet group had lost only 9.7% of their initial weight. However, the long-term results were noticeably different. After an average of 2 years (in the individual studies, this period varied between 1 and 5 years), the VLCD group had regained a great deal of their lost weight and were now only 6.3% less than their initial weight. The low-calorie-diet group was not significantly different, showing a 5% loss from their initial weight. The authors of the meta-analysis concluded that the extra risks and expense of VLCDs were not justified unless there were good medical reasons in favour of a very rapid initial loss of weight.

A similar picture has emerged from other meta-analyses and reviews of the literature on dieting. Long-term follow-ups always indicate that there is substantial regain in weight. However, analyses of these kinds, which emphasise what happens to the 'average' person in the study, hide the very considerable degree of individual variation (see Chapter 6). Within these studies, there will always be some individuals who have been able to reduce their weight in the short term and then maintain that loss over a period of many years. In Section 8.6, we shall look at the characteristics of such individuals, asking what broader adjustments in their lifestyle have allowed them to be successful.

8.4 Exercise as a strategy for reducing obesity

Exercise uses up energy and in so doing depletes the energy reserves in the skeletal muscles.

◆ What energy reserves are there in muscles?

◆ Glycogen is the energy store in muscles (see Section 2.4.1).

A fall in the level of stored muscle glycogen triggers release of glucose and fatty acids from the liver and adipose tissue and reduces the quantity of nutrients directed along biochemical routes to storage. (Recall that this is similar to events that occur during fasting – described in Section 4.3.3.) It follows that an increase in exercise, provided that energy output now exceeds energy intake, should lead to a decrease in weight, as it did for Paul (see Section 8.1). Unfortunately, there are a number of factors that complicate this otherwise simple route to weight loss, and the evidence that exercise produces weight loss in overweight people is mixed.

Physical exercise may, broadly, fall into one of two categories: aerobic and anaerobic. In **aerobic exercise**, the body is able to provide a continuing and sufficient supply of oxygen to the exercising muscle. Glycogen is completely metabolised to glucose and then via the TCA cycle to carbon dioxide and water to provide energy (see Figure 2.6). In **anaerobic exercise**, the oxygen supply is inadequate and glycogen can only be broken down as far as pyruvate, which is further metabolised to lactate, rather than to carbon dioxide and water (see Box 4.1). Exercise involving short periods of very intense activity – typical of sprinting, training with heavy weights or other types of resistance training like 'sit-ups' and 'stomach crunches' – is typically anaerobic.

8.4.1 Aerobic exercise programmes and weight loss

The great majority of studies looking at the impact of physical activity on body weight use aerobic exercise programmes. Ross and Janssen (2001) looked at 31 studies on the effect of exercise on weight change in people with a BMI above 25. They found that the greater the amount of energy expenditure by physical activity, the greater the reduction of body weight and total body fat. The results of one study (Jakicic et al., 2003) involving 201 women with BMIs above 27 are shown in Figure 8.3.

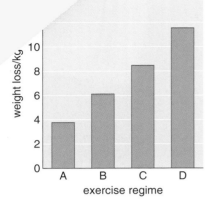

Figure 8.3 The relationship between weight loss over 12 months and amount of weekly activity. One group exercised for less than 150 minutes per week (regime A); a second group exercised for more than 150 minutes per week for the first 6 months, but less than 150 minutes per week for the second 6 months (regime B); a third group exercised for more than 150 minutes per week (regime C); a fourth group exercised for more than 200 minutes per week (regime D).

◆ Look at Figure 8.3. What is the relationship between the two variables, weight loss and amount of activity?

◆ There is a positive correlation between the two variables. As the amount of activity increases, so does weight loss.

In another study, Jeffery and colleagues (2003) compared low (1000 kcal per week) and high (2500 kcal per week) levels of physical activity. Weight losses for the two groups did not differ significantly at the end of 6 months. However, participants in the high-activity group maintained their losses significantly better at both the 12- and the 18-month follow-up assessments than did those in the low-activity group.

Ross and Janssen also suggest that there was considerable variation in body weight reductions. Differences in the amount of activity, the amount of moral support, the weight of the participants and the mood of the participants would all have a bearing on the outcome of the studies. These, then, are some of the complicating factors that impede the seemingly straightforward route from exercise to weight loss. There are others that impact on particular aspects of weight loss.

One set of complicating factors relates to the pay-off between increased exercise and noticeable weight loss. Although Paul's 9.8 kg is a respectable amount of weight to lose in a year, it was only achieved by taking daily activity, an hour in total each and every day with probably very little reward (i.e. noticeable weight loss) in the first few weeks.

A second set of complicating factors involves the extent to which *compensation* occurs. Adding an episode of exercise to a daily regime may have unexpected consequences; an extended period of compensatory inactivity (rest) may be added to the day, or energy intake may increase as a 'reward' for the exercise. The problem of such compensatory behaviour was well described by William Banting in 1869:

> I consulted an eminent surgeon, now long deceased,—a kind personal friend,—who recommended increased bodily exertion before my ordinary daily labours began, and who thought rowing an excellent plan. I had the command of a good, heavy, safe boat, lived near the river, and adopted it for a couple of hours in the early morning. It is true I gained muscular vigour, but with it a prodigious appetite, which I was compelled to indulge, and consequently increased in weight, until my kind old friend advised me to forsake the exercise.

Banting's pamphlet is usually recognised as the first English language diet book, and his response to exercise is not unique. A study was made of 35 overweight or obese individuals who participated in a structured exercise programme over a 12-week period (King et al., 2007). Measurements of body weight, BMR, appetite and food intake were made at the beginning and end of the study. When the researchers looked at the data in detail, they realised that some individuals showed no compensation whatsoever for the energy that they used in the exercise programme. As a consequence, they lost a considerable amount of weight. Other individuals, while still maintaining the required level of exercise, also showed an increase in appetite ratings and food intake. As a result, their weight loss was very much less than might have been expected from the amount of exercise taken. Several members of this group actually increased in weight, although not to the same extent as implied by Banting's remarks! It is possible that some of the individual variability in response to exercise seen in such studies relates to the gene–environment interactions described in Section 6.4.

Such compensation may be entirely inadvertent, but the consequence will be to reduce the impact of the exercise regime on body weight. (One of the main assumptions underlying the weight-loss calculations for Paul was that there were no other changes in his energy intake or energy expenditure.)

A third set of complicating factors relate to persistence. In general, individuals who are overweight or obese do not take as much exercise as do those of healthy weight.

Even small differences in BMI appear to influence activity levels. Figure 8.4 shows results from a study conducted in Cameroon.

◆ What can you say about the relationship between BMI and the amount of physical activity?

◆ There is a negative correlation between BMI and the amount of physical activity. Those men with the highest BMI are the ones who take the least amount of physical activity.

Many other studies have found that overweight and obesity are negatively correlated with the overall level of physical activity in adults. So any exercise programme for weight loss in obesity is probably trying to introduce a new kind of behaviour rather than trying to increase the levels of an existing behaviour. Indeed, it is found that drop-out rates from exercise programmes are relatively high and a high degree of overweight is one of the most consistent predictors of drop-out from exercise programmes. Reasons for such low persistence are not that hard to find. When discussing Paul, we mentioned the relatively slow pay-off and the general sense of the exercise not making any difference. To this can now be added the actual and perceived extra effort exercise entails. The extra effort arises not just from the activity itself but from the movement of heavier limbs and a heavier body. This is the reason for exercise tables giving different energy values for people of different weights: there is more body to move. To a lesser extent, large bodies are also mechanically inefficient, as described in Section 7.2.1. Hence there are straightforward mechanical reasons that make, say, walking briskly more difficult for larger people. In recognition of this, some training courses for doctors involve putting on a 'fat suit' (Figure 8.5) and experiencing at first hand the difficulties that a bulky body imposes.

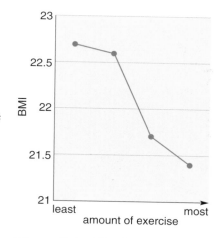

Figure 8.4 Physical activity and its relationship with BMI in men living in Cameroon.

Figure 8.5 This journalist wore a 'fat suit' for a day and experienced some ridicule and discrimination.

Figure 8.6 These men are using rowing machines to provide resistance training. Exercising with a 'buddy' can be more motivating than working alone.

8.4.2 Anaerobic exercise, BMR and weight loss

As an alternative to (or in addition to) aerobic activities, such as walking or swimming, some people push or pull against a force (Figure 8.6) or lift weights to increase strength – so called resistance training. Such training allows those with limited mobility to exercise and is considered to be anaerobic exercise if sufficient effort is put into the activity. Resistance training has one notable benefit over aerobic exercise: it increases the amount of muscle (lean tissue) in the body. Lean tissue has a higher metabolic rate than adipose tissue (Section 2.7), and so an individual who undertakes resistance training will increase their overall basal metabolic rate if the ratio of lean tissue to fat tissue increases.

There has been some controversy over the ability (or otherwise) of regular exercise to bring about a long-term increase in TEE. Some studies have found an increase in TEE after exercise; others find an increase in BMR following exercise. However, these studies have tended to look for these differences rather soon after an exercise programme has finished. In a study designed to address these issues, Poehlman and colleagues (2002) compared two exercise regimes: resistance and endurance training (anaerobic and aerobic, respectively). They recruited inactive young women, aged between 18 and 35, who had stable weight with a BMI of less than 26, and randomly allocated them to one of three groups: an aerobic activity group; a resistance training group; or a control group. Forty-eight participants completed the 6-month programme, some 54% of those starting out. There was a statistically significant increase in muscle (and muscular strength) in participants undergoing resistance training, and therefore their BMR had also significantly increased. The endurance training group had increased their aerobic capacity by 18% (i.e. they had increased their ability to deliver oxygen to the exercising tissues). Despite this increased fitness in both training groups, 10 days after the 6-month training period there were no significant or sustained changes in TEE in any group. Thus the value of the training programme had been in the direct cost of exercise and not in any long-term change in daily energy expenditure.

The authors of the report draw attention to two factors of relevance to anyone interested in exercise as a means to encourage weight loss.

1   The recruited women were not overweight, and they were eager to take part in the trial. Nevertheless, a third of them dropped out – the majority of them having no particular reason for non-compliance, once again demonstrating that it can be difficult to persuade individuals to persist with an exercise regime.

2   The authors felt that they had perhaps been naive in thinking that after a 6-month period these women would continue to exercise without the encouragement of their group and trainer.

8.4.3 Additional health benefits of exercise

Although the previous sections may suggest that exercise is not especially effective in promoting weight loss, it is important to remember that it has a number of additional health benefits. Exercise, especially aerobic exercise, increases cardiovascular fitness (e.g. the 18% increase in aerobic capacity noted above), and cardiovascular fitness is known to be a better predictor of cardiovascular health than body fat measures. Indeed, individuals who are overweight but fit are actually less likely to die from heart disease than those whose weight is normal but who are unfit. Exercise also increases glucose tolerance and insulin sensitivity (see Vignette 8.1). There is evidence to suggest that this effect is independent from any reduction in body weight.

Vignette 8.1   Ned does some gardening

Ned has gradually increased in weight throughout his adult life. He became obese after he had to take early retirement, and he then developed diabetes. Although his doctor encouraged him to lose weight and suggested a daily walk, Ned wasn't fond of walking. In the spring, he visited the doctor because he was not coping well with controlling his diabetes. During the consultation, the doctor discovered that Ned used to be a keen gardener. He suggested to Ned that now was the perfect time to get out into the garden again and that it would improve his health. Ned looked much happier when he visited his doctor in the autumn. He hadn't taken off weight, but when the results of his OGTT came, his insulin sensitivity was improved.

The results from one study that shows this effect (Nassis et al., 2005) can be seen in Figure 8.7. Nassis recruited 19 overweight and obese teenage girls, of whom 15 completed a 12-week aerobic exercise programme. They took an OGTT before and after the 12-week programme.

◆ What was the effect of the exercise programme?

◆ The exercise programme had little effect on glucose levels and the rate at which glucose was removed from the blood in the OGTT. However, the amount of insulin needed to reduce the glucose levels was lower after the exercise regime than before the exercise regime, i.e. insulin sensitivity had been increased by the exercise.

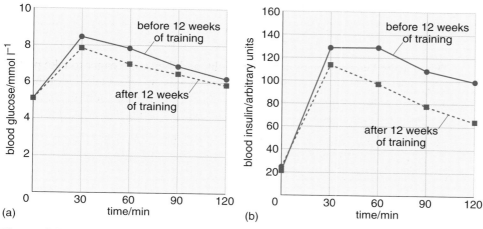

Figure 8.7   Improvement in insulin sensitivity in overweight and obese teenage girls after 12 weeks of aerobic training. (a) Blood glucose and (b) blood insulin responses during the 2-hour OGTT. Glucose was administered at time 0.

Exercise may also be associated with enhanced mood, reduced anxiety, increases in self-esteem and in cognitive function.

8.4.4 Exercise and body shape

Where do you lose fat from when exercising? Not from where you most desire. Figure 8.8 shows the availability of fatty acids during exercise.

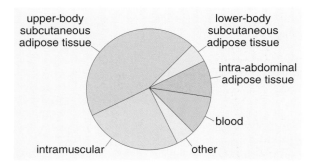

Figure 8.8 The availability of fatty acids from different body compartments during exercise.

◆ Think about the relationship between fat depots and the comorbidities associated with obesity. From which depot would fat loss deliver the greatest health benefits?

◆ The loss of visceral (intra-abdominal) fat would deliver the greatest health benefit (see Section 1.2).

As you can see, of the three storage areas – intramuscular, subcutaneous and visceral – it is the visceral storage area that makes the smallest contribution of fatty acids to the energy consuming muscles. Many women will be unsurprised by the relative contributions of the two subareas of subcutaneous fat. Fat on the hips and buttocks always seems to be immoveable!

8.5 Combining exercise and diet

It is often thought that weight loss will be increased if exercise and diet are combined. This idea seems attractive and obvious because both sides of the energy equation are modified in the right direction. However, it is sometimes suggested that the benefits of exercise and diet combined may be substantially greater than would be predicted from the effect of each one taken alone – an example of a **synergistic effect**. This section will examine some of the evidence for synergistic effects of diet and exercise in tackling the challenges of obesity.

A study by Redman and colleagues (2007) investigated whether similar changes in energy balance achieved either through diet alone or through the combination of diet and exercise would have different effects on weight loss and body composition over a 6-month period. They expected (and hoped) that the diet/exercise combination would produce a better outcome in terms of loss of visceral body fat as well as total body weight. Their study had three groups of participants, all of whom were overweight but had no additional health problems. The first group, acting as a control, was prescribed a diet that was expected to simply maintain their current weight. A second group was prescribed a diet that reduced their overall energy intake by 25%. The third group was prescribed a diet that reduced their energy intake by 12.5% and an exercise regime that would increase their energy output by a further 12.5%. Before and after the 6-month study period, the participants went through a series of measurements of the distribution of total body fat and visceral fat.

◆ What results would you expect from this study if diet and exercise had a synergistic effect on weight loss and body fat distribution?

◆ Weight loss and body fat distribution should show more beneficial changes in the combined diet/exercise group than in the diet-only group, despite the similarity of their changed energy balance.

The results of the study were clear cut, and a surprise to the investigators. Overall, the participants in the two active treatment groups lost about 10% of their body mass and about 24% of their fat mass. These differences were highly significant. However, there were no differences in either measure *between* these two treatment groups, and therefore no evidence for a synergistic effect of diet and exercise on loss of body weight in the study. They concluded that the broader beneficial effects of exercise on health in overweight and obese people are likely to be indirect, occurring as a consequence of increased cardiovascular fitness rather than on any direct effect on body weight. Although this may seem a rather negative conclusion, we have seen in the previous section that the additional benefits on both cardiovascular fitness and insulin sensitivity are worthwhile. In addition, combining exercise with a diet will help reduce the loss in lean tissue that might be expected with a more severe diet.

8.6   Lifestyle and body weight

If you have been reading this chapter in the hope of finding a solution to obesity, at either an individual or a more global level, then you are probably feeling a little discouraged. The evidence suggests that, for many individuals, neither dieting nor exercise lead to substantial long-term weight loss. Why might this be? Part of the answer lies in the difficulty that many people have in making *permanent* changes in their behaviour, even if, at a conscious level, they wish to make those changes. Being 'on a diet' carries the strong implication that, sooner or later, the diet will be abandoned. Health clubs, whose advertising emphasises the health benefits of exercise, have found from experience that many members, despite paying a regular monthly subscription, only visit infrequently and drop their membership after a year or two.

The difficulty of making permanent changes in behaviour is, of course, not restricted to eating and exercise. Cigarette smoking and excessive alcohol consumption are just two examples of other unhealthy behaviours where individuals may repeatedly try (and fail) to change. These cases are often described in terms of addiction and relapse. Alcohol and nicotine (from cigarettes) are chemicals introduced to the body, but for which the body has no requirement (unlike food). However, over time, the body can change to develop a requirement for these drugs, after which addiction is said to have developed. Once addicted, to stop taking the drug can be very uncomfortable and difficult, but after a relatively short period of time the physiological effects of withdrawal of the drug will have completely disappeared. So this factor cannot explain relapse. However, the unconscious associations made between the use of the drug, the rewarding effects that it had and the situations in which the drug was used may remain virtually intact after even a long period of abstinence. As a result, perhaps when a former user finds themselves in a situation in which they took the drug, they may again find that they crave it.

Although the exact psychological mechanisms underlying this process are not fully understood, it is reasonable to suppose that the same mechanisms may also contribute to the difficulty in cutting out highly desirable, energy-dense items from an everyday diet. The sight of attractive buns and cakes, the smell of coffee, and even the brand images from a local coffee house may powerfully arouse the appetite even of those who have successfully resisted such items for months or years. It is not surprising that food retailers are so protective of their brand images (Figure 8.9; see also Section 5.3.5).

Figure 8.9 Well-known brand images.

Vignette 8.2 Nazneen finds a support group

With the encouragement of her midwife and her desire to deliver a healthy baby, Nazneen was thrilled to find that through healthy eating she put on no further weight in the final trimester of her pregnancy. She was lucky in that both she and her baby son took readily to breastfeeding. When he lost weight after the birth, she was pleased to discover that this is exactly what is expected and is believed to be protective against childhood obesity.

A few weeks later she noticed an advertisement for a club for young mothers who want to regain their figures run by a woman who, from her name, sounded as though she might be from Bangladesh too. Nazneen rang up and found that this was indeed the case and that as a practising Muslim she also understood the difficulties of taking exercise while dressing decorously. Tariq approved of her joining the group and, with the support of her new friends, Nazneen breastfed for over 6 months – and lost weight too. The group also produced a booklet of traditional recipes that were less energy-dense than those that Nazneen had been cooking but just as delicious – so Tariq was happy and was losing weight too.

Many individuals who are about to go on a diet will think of joining a class (see Vignette 8.2). This may be organised at the workplace, sponsored by a local authority or health trust, or run by a commercial organisation such as WeightWatchers®. These classes place diet and exercise in a social or group context and have a number of features that may help to establish more permanent changes in behaviour. Early meetings of such groups will typically emphasise the negotiation and acceptance of particular goals in relation to decreased food intake and body weight and to increased exercise. Participants will be encouraged to report back on the extent to which they have maintained these goals. In addition, such groups promote the use of behavioural techniques that help to maintain adherence to a planned diet. These include:

- information on diet and nutrition so that participants find it easier to make healthier choices when shopping or eating out.

- use of **stimulus control** techniques. A stimulus in this context could be anything to do with food or eating: a picture of a roast dinner, the smell of bacon, bread in a baker's window, a packet of biscuits in a cupboard. Stimulus control requires reducing exposure to cues that are likely to increase snacking between meals, e.g. by keeping food out of sight and only eating in a particular room while at home. In addition, the use of pre-planned shopping lists will help to avoid impulse purchases of attractive but unhealthy food while shopping.

- changes in eating behaviour. Eating slowly, carefully chewing food to extend mealtimes despite smaller portions. This has the additional benefit that subsequent digestion and absorption occur more rapidly, promoting the release of satiety signals within the body (see Section 5.4). Increasing water consumption during a meal will also enhance stomach distension, adding a further satiety cue.

Specific psychological interventions may provide useful additions to these types of programme. One such intervention (Luszczynska et al., 2007) was based on techniques derived from the **theory of planned behaviour**. This influential set of ideas has been applied to a number of areas in health psychology, including effective dieting. In brief, the theory suggests that much human behaviour is guided by a number of different kinds of belief. These include beliefs about:

1   the consequences of behaviour, and the extent to which these are good or bad

2   the extent to which behaviour fits the expectations of others and is acceptable

3   the factors that are likely to either help or hinder attempts to behave in a particular way.

The first stage of Luszczynska's intervention elicited information about these different beliefs from the participants in her study. They were a group of women who had enrolled in a WeightWatchers class. In the next phase of the study, the participants explored and rehearsed short plans that summarised the way in which they would overcome points at which they might 'break' their diet.

Let us consider how this might work in relation to eating a piece of chocolate cake.

1   I often eat chocolate cake as a reward when I've finished the monthly accounts. The consequence is I feel better in the short term, but in the long term I'm putting on weight. I should plan not to eat cake but have a piece of my favourite fruit available as a reward.

2   I eat cake when it's offered in the office on someone's birthday. It is expected that I will accept it. Because it seems rude to refuse I need to plan ahead and explain about my diet in advance; then it will be acceptable to refuse.

3   As I walk round the aisles in the supermarket I'm always attracted to the chocolate cake. My plan should be to avoid the aisle marked 'cakes and biscuits'.

These short plans are known as *implementation prompts*. In Luszczynska's study, the group of individuals who used this technique were compared with a control group of women who received general support time equivalent to that received by the experimental group. After 2 months, the participants using implementation prompts had lost 4.2 kg on average, whereas participants in the control group had lost only 2.1 kg on average – a statistically significant difference.

Section 5.4.1 referred to the process of implicit learning in which we can learn the relationship between stimuli and their consequences without, necessarily, being aware of it. Again, taking the example of a chocolate cake, knowledge about the pleasant characteristics of the cake may be partly *implicit* as well as **explicit** (something we are conscious of, and can reflect on). Attempts to change

Figure 8.10 Campaigns to reduce cake consumption may increase implicit liking for cake if the images used in the advertising are attractive.

behaviour, which might involve reading the label, working out how much fat and how many calories are present in a serving, may change explicit knowledge but not affect implicit knowledge. As a consequence, when presented later with a piece of this cake, someone may say that this is something they do not want because it is unhealthy, but nevertheless have a positive implicit response. A number of studies suggest that the behavioural response to the cake (eating it, or not) is very powerfully influenced by this implicit response to it. Some kinds of educational campaigns have actually been shown, paradoxically, to strengthen these responses. Thus, a poorly designed healthy eating campaign that aims to reduce cake consumption may appear to have the right effect, in the sense that people exposed to it acknowledge that cakes are high in fat and calories and may be associated with heart disease, but may actually increase their implicit liking for cake because it looks so delicious! (Figure 8.10). The unfortunate result is that people exposed to this campaign won't eat less cake.

8.7 Childhood obesity

Childhood obesity is recognised as an especially serious problem because of the strong relationship between childhood obesity and subsequent adult obesity. The earlier obesity develops in childhood, the more serious it is likely to be for that person as an adult. In an extended series of studies, Leonard Epstein, working in New York, has developed a family-based approach to childhood obesity which has been shown to be highly effective. The technique involves both the overweight children and their often overweight parents. They are involved in educational classes in which they all learn to use a 'traffic light' system (red, yellow and green codes for the range of bad to good food choices) for choosing their diet. There are also regular classes involving exercise and attempts to reduce behaviour that promotes inactivity. Epstein comments that the average weight of children enrolling in the programme has increased in the 25 years during which these studies have been conducted. He also notes that opportunities for inactivity have become greater with the development of computer games and the increased availability of television. The availability and attractiveness of energy-dense food and drink have increased and their relative cost has decreased. However, the effectiveness of the programme has been demonstrated to be highly significant in statistical terms, and has been maintained over the period during which the studies have been conducted (Epstein et al., 1995). Similar programmes, although not so stringently validated, have been introduced in other countries, including the UK. The London-based MEND programme is one well-publicised example. The acronym MEND stands for:

- Mind: understanding and changing unhealthy attitudes and behaviours around food

- Exercise: adequate, safe and – above all – fun exercise

- Nutrition: enjoyable, practical activities that teach children about healthy eating and daily meal planning to improve the whole family's diet

- Do it!

Summary

To summarise the last two sections, a multidisciplinary approach is most likely to help both adults and children to permanently regain a healthy weight. Long-term or permanent changes in lifestyle are the key aim and are being assisted by increasingly sophisticated psychological techniques. However, at least for many individuals, our present-day obesogenic environment too readily promotes excessive calorie intake, an inactive lifestyle and an unhealthy degree of weight gain. In Chapter 10 we shall ask to what extent society at large should take responsibility for modifying the obesogenic environment, especially in the way it impacts on more vulnerable members of the population such as children.

8.8   Summary of Chapter 8

8.1   Using MET values and the energy equation reveals that the theoretical long-term weight loss following small changes in exercise or diet can be substantial.

8.2   Diets with an energy content below that required to maintain weight can be effective in reducing weight in the short term, but are less effective in the longer term because individuals return to their previous eating habits.

8.3   Diets that alter the macronutrient balance of the diet can be effective in reducing weight in the short term, but are less effective in the longer term.

8.4   Exercise on its own can lead to weight loss and has many health benefits, but the complicating factors of pay-off, compensation strategies and poor persistence often mean that any weight loss may be small.

8.5   A combined exercise and diet regime has additional health benefits, although there is probably no additional effect on weight loss beyond what each would achieve on its own.

8.6   Permanent changes to behaviour using a variety of psychological techniques – and, especially in the case of obese children, including all the family members – may be the route to permanent weight reduction.

Learning outcomes for Chapter 8

LO 8.1   Define and use, or recognise definitions and applications of, each of the terms printed in **bold** in the text.

LO 8.2   Explain the factors that influence the effectiveness of diet and exercise as treatments for an obese individual.

LO 8.3   Suggest differences between studies of weight control that might account for their differing outcomes.

LO 8.4   Understand the rationale for strategies that encourage permanent changes in lifestyle to maintain weight loss.

LO 8.5   Use calorific values and MET values to calculate potential weight loss and gain.

Self-assessment questions for Chapter 8

Question 8.1 (LO 8.2)

In what two main ways do the diets considered in this chapter differ?

Question 8.2 (LO 8.2)

What are the two main advantages of resistance exercise?

Question 8.3 (LO 8.1)

If you wanted to lose weight by regular exercise and were identified as a 'compensator', would you be pleased or annoyed? Give your reasons.

Question 8.4 (LO 8.3)

Give three reasons why one study of the effectiveness of a new diet might have a different outcome from another study of the same diet.

Question 8.5 (LO 8.4)

Why might implicit knowledge undermine an intensive education campaign to combat unhealthy eating?

Question 8.6 (LO 8.5)

Paul discovers he has put on 9.6 g per week over the past year. How many kilocalories extra has he consumed than used over the year? (Assume that Paul's weight gain is fat.)

CHAPTER 9

TREATING OBESITY: DRUGS AND SURGERY

Slim Bomb weight loss pills … For EXTREME RESULTS – without dieting!

Advertisement

9.1   Introduction

Although diet and exercise as part of a broad programme of permanent behaviour change are the appropriate response to deal with problems of excess weight for many individuals, for a minority it may be necessary to consider additional forms of treatment. This chapter examines two different types of therapy for more intractable or life-threatening cases of obesity. The first of these involves the use of pharmaceutical drugs to modify different aspects of physiology or behaviour. The rationale for different drug treatments can be understood in terms of the material found in Chapters 2–5, and we shall review some of the evidence that suggests drug treatment is sometimes a useful option for the treatment of obesity. There are also a number of surgical treatments for obesity, and the final section of this chapter will examine some of the possibilities in this area. Drugs and surgery have a wide variety of potential undesirable side effects. It is therefore important, using evidence from appropriately designed clinical trials, for clinicians and their patients to evaluate whether the gains of treatment are outweighed by harmful consequences.

9.2   Drug treatments for obesity

9.2.1   When is drug treatment appropriate?

The clinical guidelines for obesity treatment from the UK's National Institute for Health and Clinical Excellence (NICE, 2006) suggest that drugs should be considered:

- only after dietary and exercise advice have been started and evaluated by health professionals
- for patients who have not reached their target weight or have reached a plateau in weight reduction.

At present (2008), NICE approves two drugs, sibutramine and orlistat, and offers the following comments:

> Current treatment of obesity may include dietary and lifestyle advice, and/or pharmacological treatments, such as anti-obesity drugs orlistat and sibutramine. Published NICE guidance recommends that orlistat should be prescribed for the management of obesity in people with a BMI of 30 kg m^{-2} or more, and in people with a BMI of 28 kg m^{-2} or more and significant comorbidities. The guidance recommends that sibutramine should be prescribed for the management of obesity in people with a BMI of 30 kg m^{-2} or more, and in people with a BMI of 27 kg m^{-2} or more and significant comorbidities. Sibutramine is contraindicated in people with a history of cardiovascular disease.

◈ What is meant by an individual having a 'BMI of 28 kg m^{-2} or more and significant comorbidities'?

◆ The individual is overweight (BMI exceeds 28) and also suffers from additional medical conditions which are harmful to their health and likely to be made worse by their excess body fat. Such medical conditions might include diabetes or cardiovascular disease.

Thus sibutramine or orlistat may be used even for individuals who are at the high end of the 'overweight' classification, although the guidance suggests that it is inadvisable to prescribe sibutramine ('is contraindicated') to people who have cardiovascular disease. One further drug, rimonabant, has been approved by European drug regulators, but NICE have yet (as of early 2008) to publish their guidance on this compound.

9.2.2 Current options for the pharmacological treatment of obesity

Orlistat

Orlistat is the only approved drug whose mechanism of action is to reduce the absorption of ingested nutrients. It acts by inhibiting the digestion of fats.

◈ How are fats digested in the gut?

◆ They are broken down into simpler molecules before being absorbed across the gut wall and entering the circulation.

Fats are made up of triacylglycerol molecules and are broken down into two free fatty acids and monoacylglycerol before they can be absorbed across the gut wall. This process is carried out by the enzyme known as lipase (Section 3.3.3). Orlistat, whose trade name is Xenical® in much of the world and alli™ in North America, is able to inhibit the action of lipase and thus reduce the absorption of fat.

When orlistat is taken at the prescribed dose, it inhibits the absorption of about 30% of ingested fat. One consequence of the reduced level of fat breakdown and absorption may be a decrease in the postprandial release of CCK in response to a meal (Section 5.4.3).

◈ What effect might the reduction in CCK release have on eating behaviour?

◆ It might reduce the satiety effect of CCK, which would not be helpful in reducing food intake.

There is good evidence to suggest that orlistat can promote weight loss. The clinical trials used to test the effectiveness of the drug were similar in design to those discussed in Chapter 8. Patients – in this case, people who were clinically obese – were randomly assigned into two or more groups. They then received either active drug (different groups may receive different drug concentrations) or a **placebo** treatment (the control condition). A placebo is an inert drug treatment which, to the patient, appears identical to the active treatment, and ideally neither the patient nor the prescribing clinician will know which patients receive the active drug treatment. Not knowing who is receiving what, i.e. being 'blind' to

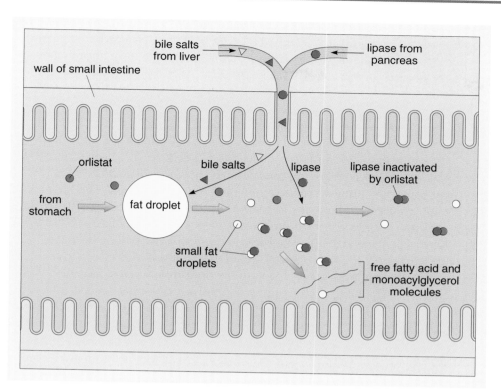

Figure 9.1 Schematic diagram showing the interaction between orlistat and lipase.

the treatment condition, avoids bias in the assessment of the patients and allows estimation of the extent to which the expectation of receiving an active treatment can modify the patients' condition. Trials of this kind are described as **randomised double-blind trials**; *randomised* because patients are randomly allocated to treatment groups and *double-blind* because neither the patients nor the treating clinician know which treatment each individual is receiving. Other clinicians who are not directly treating patients will continually review the incoming data and will intervene to stop the trial if significant side effects of the active treatment begin to emerge. In practice, however, both patients and their clinicians may gradually become aware of whether they are receiving active drug or placebo.

Figure 9.2 shows the changes in weight recorded over 4 years of treatment with orlistat or placebo. Other data from this trial showed worthwhile reductions in fasting insulin and glucose levels. In addition, obese patients were less likely to develop diabetes if they were in the active drug treatment group.

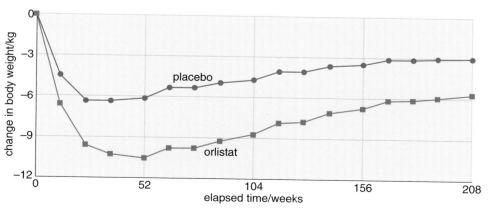

Figure 9.2 Weight changes following 4 years' treatment with orlistat or placebo.

Clinical trials and experience have suggested that orlistat is a relatively safe drug. In part this is because it is not absorbed into the body but remains in the gut until it is eliminated with the faeces. Therefore, it is less likely to have undesirable side effects on essential organs such as the heart, lungs or brain. Such problems, as you will discover later in this section, have plagued many other pharmacological treatments of obesity. However, this does not mean that orlistat is free of side effects. The failure to absorb fat means that the fat remains in the gut and can alter the properties of the faeces. They may become oily and loose and, in some cases, this can lead to faecal incontinence. The US distributor's website suggests 'Until you have a sense of any treatment effects, it's probably a smart idea to wear dark pants, and bring a change of clothes with you to work.' The fat may also be digested by bacteria in the lower gut, leading to flatulence (Section 3.2). Again, the website advice is specific: 'You may not usually get gassy, but it's a possibility when you take alli. The bathroom is really the best place to go when that happens.' Of course, as you would expect, these side effects can be minimised by reducing fat intake. Users of the drug are normally provided with advice that will help them to adopt a low-fat diet, and this expected side effect of treatment probably 'helps' them comply with this advice as well as reducing their energy absorption.

Sibutramine

Sibutramine, which is marketed under the name of Meridia™, has a very different mode of action from orlistat. Its primary effect is to reduce appetite by acting on brain neurotransmitter systems; however, it may have an additional effect of raising BMR, which will be discussed later in this section. The drug was originally developed as an antidepressant but was not found to be effective enough to justify its clinical use. However, those participating in the trials did show reduction of body weight. Sibutramine modifies the functioning of brain synapses that use serotonin (also known as 5-hydroxytryptamine or 5-HT) as a neurotransmitter. Under normal circumstances, the neurotransmitter is released, acts at receptors and is then deactivated by being taken back into the cell that released it (Figure 9.3). Sibutramine is an inhibitor of this re-uptake process. The effects of sibutramine are not restricted to synapses where serotonin is the neurotransmitter; sibutramine also inhibits the re-uptake process at synapses that use noradrenalin as their neurotransmitter.

◆ Serotonin and 5HT are two names for one chemical substance. It sometimes happens, as in this case, that scientists in different laboratories are working independently on a chemical that they have found in the body. They might have located this chemical in different areas and investigated different aspects of its activity. They name the chemical and publish their findings. It may be some time before anyone realises that the two names (and all the associated findings) relate to a single substance.

◈ Looking at Figure 9.3, what would you expect to be the consequence of reducing neurotransmitter re-uptake on the functioning of the synapse?

◆ It will increase the effect of the neurotransmitter on the postsynaptic neuron because the neurotransmitter will remain in the synapse for longer, becoming more concentrated and thus having a more prolonged effect at the receptors.

It is thought that the most important action of sibutramine in relation to eating is to enhance the action of serotonin within the brain, and in particular within the arcuate nucleus of the hypothalamus (Section 5.5.1). Serotonin receptors are expressed on the αMSH-containing cells within this area. Activation of these receptors is known to increase the activity of the αMSH-containing cells, thus enhancing their inhibitory effect on feeding behaviour (Figure 5.13).

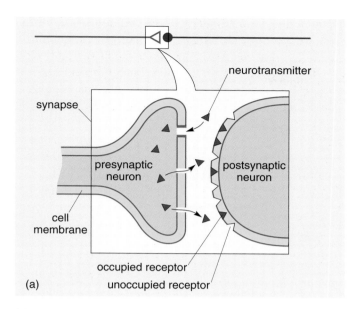
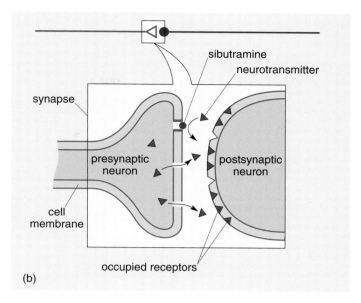

Figure 9.3 (a) The action of a neurotransmitter at a synapse under normal conditions. (b) Sibutramine blocking the re-uptake mechanism.

When we discussed the effects of neurotransmitters and hormones at their receptors (Chapters 1 and 5; see Figures 1.7 and 5.6), we did not mention one important aspect of these systems. You might imagine that for each neurotransmitter there is a single matching receptor. In fact, there may be several different types of receptor for a given neurotransmitter. These are referred to as *receptor subtypes*. Almost all neurotransmitters are known to have more than one receptor subtype, and serotonin has as many as 14.

◆ Recall that neurotransmitter receptors are large protein molecules inserted into the neuron's cell membrane. What does this imply about the genetic coding for neurotransmitter receptors?

◆ There must be separate genes coding for each of the receptor subtypes.

It is well known that different genes may be expressed in different tissues; they become active, sending messenger molecules into the cytosol, where they are used to manufacture the relevant protein (Figure 2.5). That is why, for example, a cell that lines a blood vessel is different from a neuron. The genes that code for receptor subtypes may also be expressed in the neurons that develop in one area of the brain, but not in another area that has a different function.

Many drugs act as mimics of neurotransmitters. This means that the three-dimensional structure of the drug shares enough characteristics with the neurotransmitter to allow it to bind at the same site on the receptor as the neurotransmitter. However, the different receptor subtypes for a given neurotransmitter will frequently bind different parts of the neurotransmitter molecule. This principle is shown, in a very simplified form, in Figure 9.4. This allows chemists to design drug molecules which act as a mimic of the neurotransmitter at one receptor subtype, but not at another.

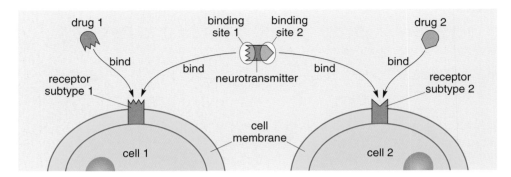

Figure 9.4   A neurotransmitter can affect two different cell types, cell 1 and cell 2. It has two binding sites: binding site 1 binds to receptor subtype 1 and binding site 2 binds to receptor subtype 2. Drug 1 interferes with the neurotransmitter's activity at receptor subtype 1, but not at receptor subtype 2, thus affecting neurotransmitter activity only in relation to cell 1. Drug 2 has the opposite effect.

When a drug acts as a mimic of a neurotransmitter (or hormone) at a receptor, it may have one of several distinct effects. In the simplest case, it will have the same effect as the neurotransmitter. So, if the binding of the neurotransmitter either activates or inhibits neural activity, then the drug does the same as the neurotransmitter.

◆   How would you describe a drug that has the same effect as a neurotransmitter?

◆   A drug of this kind is known as an agonist (Section 5.3.3).

If the agonist is highly effective at one receptor subtype and almost entirely ineffective at other receptor subtypes, then it is referred to as a **selective agonist**. However, many agonists, especially those whose chemical structure resembles the natural neurotransmitter, will be **non-selective agonists**. Other drugs may have different effects when they bind to the receptor. Rather than mimicking the effect of the natural neurotransmitter at the receptor, they may instead simply bind to the receptor but have no further effect. Such drugs, especially if they bind strongly and persistently, will also stop the relevant neurotransmitter from having its usual effect. You may recall that such drugs are referred to as antagonists. As with agonists, they may be more selective or less selective.

◆   Could sibutramine be referred to as a receptor agonist at serotonin (5-HT) receptors? (Look carefully at Figure 9.3 to identify the site of sibutramine action.)

◆   No, sibutramine does not act at particular serotonin *receptors*; it simply increases the availability of serotonin within the synapse by blocking the re-uptake mechanism.

Although sibutramine is not a serotonin agonist, you will occasionally find it referred to as an **indirect agonist** because the effect of enhanced serotonin availability on neurons expressing serotonin receptors will be similar to that of an agonist drug that acts directly at those receptors.

◆ Is sibutramine likely to have selective effects on synapses that use a particular serotonin receptor subtype?

◆ No, sibutramine blocks the re-uptake mechanism; it does not act at particular types of serotonin receptor.

A little later in this chapter, we shall look at the way in which more selective drugs than sibutramine may, in the future, be developed as treatments for obesity. In other therapeutic areas, this type of development has already successfully taken place. For example, there is a specific serotonin receptor subtype, known as the 5-HT$_3$ receptor, that is expressed in brain areas controlling the vomiting response. Selective 5-HT$_3$ antagonists are now widely used to reduce both actual vomiting and the feeling of sickness that is associated with chemotherapy for cancer. As a consequence, patients are able to tolerate more aggressive treatment regimes, and their survival is improved.

We turn now to consider the effect that sibutramine may have on BMR. In addition to the effects of sibutramine at synapses using serotonin as a neurotransmitter, the drug also inhibits the re-uptake of the neurotransmitter noradrenalin. It is believed that this action underlies a second mechanism through which sibutramine may help to reduce body weight. Two studies have helped to identify an effect of sibutramine of increasing metabolic rate – effects of this type are sometimes described as *thermogenic* because their effect is to raise body temperature and increase energy use through loss of heat. In the first study, the participants received a single dose of sibutramine (Hansen et al., 1998). A little while later, they reported a reduction in hunger, measured using standard rating scales – similar to those used in the experiment described for Figure 5.8. In addition, a technique known as indirect calorimetry was used to measure the metabolic rate of the participants. This procedure involves accurate measurement of oxygen use and carbon dioxide production and can even be carried out in non-laboratory environments (Figure 9.5).

◆ How, in general terms, will the measurement of oxygen use and carbon dioxide production allow the measurement of metabolic rate?

◆ Oxygen is required – and carbon dioxide is produced – by the basic biochemical processes of metabolism (i.e. it is a measure of the rate of cell respiration; see Figure 2.6). Measurement of amounts will provide an estimate of the rate of metabolism.

In this study, sibutramine produced both a decrease in hunger and an increase in metabolic rate, which was associated with increased heart rate, increased blood pressure and increased levels of adrenalin in the bloodstream. Adrenalin is a hormone that is very closely related to the neurotransmitter noradrenalin. Raised levels of adrenalin are one component of the response to stress.

Figure 9.5 Exercising while wearing an 'oxycon mobile' to measure respiratory rate.

In a second study, the same team examined the effect of sibutramine treatment over a period of 8 weeks (Hansen et al., 1999). The effects were compared with obese individuals who received a placebo treatment. Although the participants were instructed not to change their eating habits, the sibutramine-treated group lost significantly more weight than the placebo group. Measurements of energy expenditure demonstrated a reduction in TEE that was proportional to the loss of body weight. Remember from Section 2.7 that a reduction in food intake is associated with a lowering of metabolic rate that would help to conserve energy at times of food shortage. However, for those treated with sibutramine, this reduction in metabolic rate was smaller than observed in the placebo group. The effects on hunger and satiety reported in the first study (above) were replicated in the final phase of this longer-term experiment, indicating that the participants had not become *tolerant* to the drug (i.e. the drug was still having the same effect throughout the experiment).

Serotonin and noradrenalin are involved in a wide range of other functions in the body. Within the brain, serotonin is involved in the regulation of mood and noradrenalin is implicated in the processes of attention and memory. In addition, noradrenalin is involved in the regulation of heart function, insulin release and pain perception. Serotonin's other functions include the modulation of peristalsis (Figure 3.6) and also a less well understood role in lung function. Because sibutramine will affect the serotonin and noradrenalin re-uptake mechanisms throughout the body, it is not surprising that its use is associated with a number of side effects. The most serious of these relate to the effects on heart rate and blood pressure, and this explains why the drug should not be prescribed to individuals with a history of heart disease. A little later in this chapter, we shall look at ways in which neurotransmitter systems such as serotonin and noradrenalin might be targeted in a more selective way, thus avoiding these types of side effect.

Rimonabant

Rimonabant is the most recent drug to be approved for the treatment of obesity in Europe, although in June 2007 an advisory panel to the Federal Drugs Agency (FDA) recommended that the drug should not yet be approved in the USA because of concerns that it may be associated with depressed mood and suicidal thoughts. After examining the mode of action of rimonabant, we shall take the opportunity to describe the licensing process for drugs of this type and also the design of trials that test both their effectiveness and likely side effects.

Rimonabant acts on the cannabinoid system of neurotransmitters and receptors. As with several other neurotransmitters, the first evidence for its existence came from the action of a plant product on human behaviour. Various extracts of a strain of hemp plant, *Cannabis sativa* (Figure 9.6), have been used for their effect on the mind since human prehistory. Since the 1990s, very rapid progress has been made in understanding how the main active component of these extracts, Δ^9-tetrahydrocannabinol (THC), affects neural function. Although THC is not produced in the brain, a range of other compounds that act at cannabinoid receptors are produced by neurons. THC binds to cannabinoid receptors and influences neuronal activity in much the same way as these other natural neurotransmitters.

Figure 9.6 Hemp plant, *Cannabis sativa.*

◆ How would you describe THC?

◆ THC is a cannabinoid agonist. It mimics the effect of the natural cannabinoid neurotransmitters at cannabinoid receptors.

Cannabinoid receptors are located in a wide range of brain areas and have been implicated in several functions, including the regulation of mood and the formation and retrieval of memories. One effect of THC which is well known to regular cannabis users is often described as 'the munchies'. These individuals may find that their appetite is greatly increased when they use cannabis and that this is especially true for rewarding (e.g. high-fat and sweet) foods.

◆ Where, among other places in the brain, would this suggest that cannabinoid receptors might be located?

◆ In structures that are already implicated in the control of feeding behaviour such as the hypothalamus.

Cannabinoid receptors are found in both the hypothalamus and an area known as the nucleus accumbens. The hypothalamus is especially important because of its sensitivity to circulating leptin and insulin levels, and the nucleus accumbens is an important part of the brain's 'reward' circuitry that was described in Section 5.5.3.

Drugs that act as cannabinoid receptor antagonists have been developed as potential treatments for obesity. As described previously, the antagonist is able to bind to, but not activate, the cannabinoid receptor. This will reduce the effect of naturally produced cannabinoid neurotransmitters within the brain. Initial studies in rats and mice suggested that administration of these drugs can reduce feeding behaviour in the short term and also reduce body weight in the longer term. Rimonabant itself has now been studied in several large-scale clinical trials, and at least two drugs with a similar mechanism of action are undergoing current trials.

Studies of any new drug are organised within three phases. The initial Phase I studies use a small number of healthy individuals to study different properties of a novel drug, including such questions as the rate at which it is absorbed and then eliminated from the body, and its general physiological and behavioural effects. Phase II studies typically involve both healthy individuals and patients who have the relevant condition. They provide a preliminary assessment of whether the drug is likely to be an effective treatment, as well as addressing further safety-related issues. Phase III studies are large-scale trials that provide much better estimates of both beneficial actions and undesirable side effects. Only a small proportion of drugs that enter Phase I will eventually progress to Phase III, with the majority being discarded because of a lack of efficacy or because of potentially harmful side effects.

The Phase III clinical trials for rimonabant were carried out in both Europe and North America, although different experimental designs were used. However, both trials were double-blind.

In the European trial, 1507 patients were allocated to either placebo or one of two different doses of rimonabant and then remained on this treatment for the entire study period. By contrast, the North American study, involving 3040 individuals, used a 'crossover' design whereby some participants moved from one treatment to another during the course of the study. Figure 9.7 shows part of the data for the weight changes that resulted from drug treatment in the crossover study.

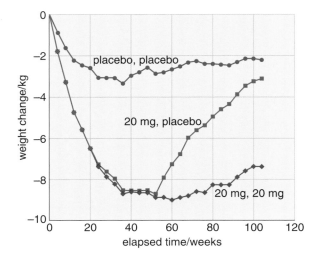

◆ Examine the group that received the drug for 1 year and then switched to placebo. What are the two most noteworthy features of their pattern of weight change?

◆ In the first year of treatment, they lost about 6 kg more weight than the group taking placebo. However, after drug treatment had been discontinued for a year, they put on weight to the extent that they are hardly different from placebo by the end of the study period.

Figure 9.7 Data for weight change over 2 years for those who took 20 mg rimonabant for 1 year and then crossed over to placebo; those who took 20 mg rimonabant for both years; and those who took a placebo for both years.

Although it is often hoped that drug treatment for obesity will help individuals to make permanent changes in lifestyle leading to maintained weight loss, the results shown in Figure 9.7 are typical of a number of other drugs. The implication is that short-term treatment with drugs that reduce appetite may not be effective in promoting permanent behaviour change and a longer-term reduction in weight. Intermittent use of such drugs may also lead to weight cycling, which can have adverse consequences and may be associated with the development of eating disorders (see Box 7.1).

9.2.3 Formerly used pharmacological treatments for obesity

A wide range of drugs have been used in the past to treat obesity. In the 1950s, amphetamine was commonly used. This drug is a powerful psychostimulant, producing a strong enhancement of mood and feelings of wellbeing. However, repeated use, especially of higher doses, may lead to dependence and addiction. Amphetamine acts on the nervous system to release the neurotransmitters dopamine and noradrenalin. A number of other drugs have a similar functional effect. For example, cocaine inhibits the re-uptake of dopamine and noradrenalin, leading to comparable psychological changes. Caffeine is another example of a (much weaker) psychostimulant. Although amphetamine does lead to a reduction in body weight, its use was soon abandoned in conventional medical circles, although anecdotal evidence suggests that both amphetamine and cocaine are still used illicitly in this way.

The reluctance to use amphetamine led to attempts to develop a drug that would still reduce appetite but not be liable to induce dependence and abuse. Fenfluramine was chemically derived from amphetamine but it has little effect on dopamine systems, instead promoting the release of serotonin, and thus it is an effective appetite suppressant. It was widely used in Europe for several decades, but substantial problems arose when it was introduced into the US market. There it was widely prescribed in combination with an over-the-counter appetite suppressant known as phentermine. Fen–phen was hailed as the miracle diet pill combination! Unfortunately, the 'miracle' soon turned sour for some of those taking the drugs. They developed a problem with their heart valve function known as 'mitral valve regurgitation', which is a complex way of saying that their heart valves failed to close properly, allowing backflow of blood as the heart contracted. The problem was not one that easily reversed when drug treatment was stopped because it had resulted from abnormal tissue growth within the valve. You might wonder how a drug that reduces appetite can also have such an effect. The answer lies in the widely distributed functions of serotonin and its receptor subtypes mentioned in the previous section. First, recall that serotonin and 5-HT are rather confusingly the identical chemical compound and that 14 different receptor subtypes for 5-HT have already been identified.

◆ What is the meaning of the statement that there are 14 different receptor subtypes for serotonin?

◆ There are 14 different genes that code for different serotonin receptor proteins.

◆ How would a selective agonist act at these receptor subtypes?

◆ Selective agonists will be able to activate one – or perhaps several – of these receptor subtypes, but will have no effect at the remaining receptor subtypes.

One of these serotonin subtypes, known as 5-HT_{2C}, is expressed in the brain, where one of its roles is in mediating satiety. A different subtype, known as 5-HT_{2B}, is expressed in the heart valves, where it can have *mitogenic* effects: effects that can promote tissue growth, sometimes in abnormal ways. Further serotonin receptor subtypes, known as 5-HT_{1B} and 5-HT_{1D}, are expressed in lung tissue, and their indirect stimulation by fenfluramine may mediate yet another harmful side effect of treatment that affects lung function. The wide range of functions of serotonin in the body makes it unsurprising that a drug like fenfluramine has substantial side effects. However, in the next section we shall see how an understanding of the specific receptor mechanism for effects on appetite allows, at least potentially, a more selective approach to drug treatment.

9.2.4 Drug treatments for obesity: future possibilities

In Chapter 5 we discussed a wide variety of hormonal and brain neurotransmitter systems that influence eating behaviour. It is not surprising that a number of these have been the subject of interest from the pharmaceutical industry. In this section, we shall look at a few of the drugs that are known to be under development. However, before we do this, it is worth thinking about the features needed in a drug for it to receive approval for clinical use.

◆ Take a few minutes to list the criteria that you think might mark out a potentially successful drug development programme aimed at one of the hormone or neurotransmitter systems mentioned in Chapter 5.

◆ Your list might include ideas such as:

- Is it possible to make a drug that acts selectively at the receptors for the hormone or neurotransmitter?

- Is this hormone or neurotransmitter involved in functions other than eating, and will this give rise to side effects of drug treatment that might be dangerous?

- Will a drug acting on this system produce effects on body weight that are at least as great as those for drugs that have already been approved?

- Could it be addictive? Are there other potential side effects?

You may well have thought of additional criteria to the suggestions above, including economic considerations such as whether the drug would be profitable, but those listed above are some of the most important from a scientific and clinical perspective. We will now examine four potential targets in the light of this discussion. Remember that drugs may, in broad terms, have two possible actions at a receptor. They may mimic the action of the natural neurotransmitter or hormone, in which case they are referred to as *agonists*, or they may block the effect of the natural neurotransmitter or hormone and are referred to as *antagonists*.

Leptin and leptin agonists

The discovery in 1994 of the gene for leptin was rightly hailed as a great scientific triumph. A lack of its product, leptin, leads not just to massive obesity, but also to a whole host of additional effects on growth and reproductive processes. It was natural to imagine that this scientific breakthrough would translate into clinical gains in the treatment of obesity. In fact, to date, this has not been the case, with the exception of a very small number of leptin-deficient individuals. Although there are no small drug molecules mimicking the effects of leptin that can be given to humans, it is possible to manufacture large quantities of leptin itself. Leptin is manufactured by inserting the human leptin gene into a bacterium and then culturing these bacteria in large fermentation flasks. Finally, the leptin can be extracted and purified into a form suitable for injection.

◈ Why inject leptin instead of taking it as a tablet?

◆ An injection has to be used because leptin is a protein (genes make proteins; see Figure 2.5). If a protein is taken by mouth, it is rapidly digested to its constituent amino acids by the enzyme pepsin in the acid environment of the stomach (Section 3.3.3).

In the clinical trials that have been reported so far, leptin has to be injected several times a day, in relatively large doses, in order to obtain even mild weight loss by comparison with placebo treatment. In one study (Zelissen et al., 2005) the obese participants were given leptin in the morning or evening only, or, for a third group, at both times. The intention had been to see whether an evening leptin injection would be more effective, because it is at this time that the levels are normally higher. Unfortunately, none of the treatments had any greater effect on body weight loss than the placebo.

These studies suggest that leptin is unlikely to be of use as a pharmacological treatment for obesity. The lack of a small drug molecule that can be taken orally and the insignificant effects of leptin on body weight mean that this strategy looks to be a poor prospect. At a scientific level, the lack of effect remains a real puzzle. As you saw in Section 5.4.3, loss of leptin function leads to massive obesity, but in most individuals there is a very close correlation between their amount of adipose tissue and the level of circulating leptin. Therefore, most obese individuals are not responding to leptin – i.e. they are *leptin resistant*. Clearly, if we understood how such resistance came about and could reverse the process, the prospects for leptin as an obesity treatment would be improved.

Recall that resistance to insulin develops in many overweight and obese individuals.

Serotonin agonists at the 5-HT$_{2C}$ receptor

As just discussed, the reasons for the withdrawal of the serotonin releaser fenfluramine as an obesity treatment was not because the drug did not work, but because its use was shown to produce potentially dangerous side effects, especially when combined with other appetite-suppressing drugs such as phentermine. However, the identification of the 5-HT$_{2C}$ receptor as the critical receptor for the effect of fenfluramine on eating prompted a number of drug companies to develop chemicals that would act specifically at this receptor. So far, the studies confirming this role have mostly been carried out in mice and rats. In addition, it has proved difficult to develop drugs that are really selective between the 5-HT$_{2C}$ receptor and the 5-HT$_{2B}$ receptor which is found in the heart. Nevertheless, the first of these drugs, developed by a US company (Arena Pharmaceuticals), has successfully completed small-scale trials in humans and does not seem to be associated with undesirable side effects. At least as important, it produces weight loss that is as great as that shown for rimonabant in Figure 9.7. The drug is now being studied in several large-scale (Phase III) trials that are likely to report in 2009–2010.

αMSH agonists

In Section 5.5.1 you examined the role of two different peptides produced in the hypothalamus that affect feeding behaviour: αMSH inhibits feeding, whereas NPY stimulates it.

◈ From the perspective of treating obesity, what kinds of drugs affecting either αMSH or NPY receptors would be of interest?

◆ αMSH agonists should mimic the effect of αMSH itself, and reduce appetite. NPY antagonists should block the effect of NPY and also reduce appetite.

Both possibilities are under active investigation. Currently (2008), αMSH agonists are showing greater promise. Drugs have been developed that selectively bind to the αMSH receptors that are found in the brain and have been subject to small-scale trials.

Bremelanotide is the name of the most advanced of these drugs. However, these studies have uncovered an additional effect of αMSH agonists that had not been expected. They induce sexual arousal in both men and women. The companies concerned (AstraZeneca and Palatin) are now investigating this as an additional potential use for the drug, and it is unclear which, if either, will be successful.

Ghrelin antagonists and CCK agonists

The functions of the gut hormones ghrelin and CCK were discussed in Chapter 5. Ghrelin, a hormone secreted by the stomach, may stimulate appetite and some studies suggest that ghrelin antagonists should be able to reduce food intake. Initially, rats were given a drug that was a ghrelin antagonist and they ate less and lost weight. These drugs also appear to promote insulin release. It will be very interesting to see whether, over the next few years, this type of drug is also effective in humans. CCK, and its role in satiety, was understood before the discovery of ghrelin. It was suggested that drugs mimicking the effect of CCK (CCK agonists) would enhance satiety and be useful anti-obesity drugs. There was some concern that the chronic administration of these drugs would disturb gall-bladder function, because CCK stimulates the release of bile from this organ. However, it is still possible that this potential use of CCK agonists will be pursued in the future.

9.2.5 Diet supplements and the treatment of obesity

Many people complain that they receive 'spam' emails offering products that can produce 'miracle weight loss'. In this section, we will look briefly at some of these products and their supposed mechanisms of action.

Figure 9.8   *Hoodia gordonii.*

Anatrim

This product is supposedly based on extracts from *Hoodia gordonii* (Figure 9.8), a fleshy-leaved succulent found in the deserts of southern Africa, and is marketed as an appetite suppressant called Anatrim. At the time of writing (2008), there is no good evidence to support this statement. Although *Hoodia* extracts are widely advertised on the internet, the actual manufacturers are often unclear. An obvious danger in purchasing such products is that you cannot be sure that what you purchase is safe to use and/or contains any active ingredient. The initial impetus for the use of *Hoodia* had come from studies in South Africa, followed by a collaboration with a UK drug company, Phytopharm. Phytopharm and the multinational pharmaceutical company Pfizer had a short-lived collaboration to identify the active principle in *Hoodia*. Phytopharm are now collaborating with the multinational food company Unilever to develop *Hoodia* extracts as a diet supplement.

Carb blockers

Carb blockers are a group of diet supplements that supposedly block the digestion of carbohydrates. The strongest claims have been made for extracts from the white kidney bean, *Phaseolus vulgaris*. There is limited evidence to suggest that, in high concentrations, these extracts may inhibit one of the digestive enzymes, alpha-amylase, that is responsible for the breakdown of starches to monosaccharides in the gut. If the action of this enzyme is reduced, then fewer of the calories from carbohydrates should be absorbed and higher levels of undigested starch will be found in the lower parts of the gut. Two very small clinical trials have been published that demonstrate a short-term reduction of weight in groups of highly selected volunteers taking very large amounts of the extract.

Effortless dieting

There is a wide range of other products that claim, variously, to enhance fat metabolism, reduce appetite or act in other ways to reduce body weight. The advertisements, such as the one at the beginning of this chapter, have some striking features in common, including claims that they reduce weight 'without effort', that they are entirely safe and that scientific research has demonstrated their mechanism of action. These claims are unsupported and you will have noticed that this advertising promotes a message that runs counter to that of the last chapter. Consistent reductions in body weight require significant engagement from an individual with a determination to produce permanent change in their patterns of energy intake and use.

9.2.6 Summary

In this section we have reviewed a number of past, present and potential drug treatments for obesity. Present treatments either reduce the absorption of fat (orlistat) or depress brain appetite systems (sibutramine and rimonabant). These treatments, as well as those such as fenfluramine that were used in the past, share some common characteristics. They produce relatively moderate degrees of weight loss. Figures of between 3 and 6 kg in addition to any placebo effect would be typical over a 6–12 month period. In addition, weight loss is not maintained after drug treatment ceases, and this is likely to be psychologically distressing for those taking the drug. Finally, the available drugs have side effects that are at least unpleasant, and sometimes more harmful. Although our scientific knowledge of the mechanisms that control eating behaviour has advanced tremendously in the last two decades, this has yet to provide real clinical benefit.

9.3 Surgical treatment of obesity

The most dramatic decision an individual and their medical advisers can make, in the attempt to lose weight, is to resort to surgery. The extent of the emotion is captured in headlines such as 'Supersize surgery saved my life' (Figure 9.9). Stories beneath such headlines often suggest that the individual concerned felt in imminent danger of death on account of their weight, but give little sense that surgery is itself an intrusive procedure carrying its own risks. Additionally, all surgical procedures have risks associated with them surrounding the use of anaesthesia and the possibility of post-surgical infections.

Figure 9.9 Some headlines from popular press and magazine articles relating to personal stories from individuals who have opted for surgical treatment of obesity.

◆ Are these risks any greater for an obese person?

◆ Yes, the relative risks associated with anaesthesia can be twice as great for an obese person as they are for an individual of healthy weight (Table 1.2, Section 1.4.2).

◆ Why is this so?

◆ There are a variety of potential complications. Airway obstruction is more likely because of fat within the neck. Cardiovascular problems are also more likely. Finally, dosing with anaesthetic drugs and gases is less reliable when there is a large amount of fat tissue in the body. This may absorb the drugs in an unpredictable manner.

In addition to the usual risks associated with surgery, there are other potentially harmful side effects linked to the particular surgical procedures that are used to treat obesity, as will be outlined later in this section. Despite this, there is evidence that, for some people, the benefits of surgery may outweigh the disadvantages.

9.3.1 When is surgical treatment appropriate?

In the UK, NICE guidelines formulated in 2006 suggest that surgery should be considered when all of the following conditions are met:

- The person has a BMI of 40 or a BMI of 35 to 40 plus other significant disease that could be improved with weight loss

- Non-surgical measures have been tried for at least 6 months but failed to achieve or maintain clinically beneficial weight loss

- The person has been receiving or will receive intensive management, including psychological support, in a specialist obesity service.

◆ Do these guidelines present any difficulties for an obese individual wishing to use surgery to reduce weight? (See Chapter 7.)

◆ There are few specialist obesity clinics (only nine in the UK in 2008), so many obese individuals will be unable to fulfil the last of these conditions.

In fact, only one-third of individuals who have surgery to treat obesity in the UK are treated by the NHS. The other two-thirds arrange and pay for their own treatment – often by taking out loans or perhaps even selling their story to the media (see Figure 9.9). Although the surgeons performing these operations in a private health care setting will probably adopt the NICE guidelines when they make their decision to accept patients, it is less easy for the surgeons to be confident that each patient will adhere to the guidelines in the future. For example, it might be that post-operative psychological support will be available to but not used by the patient.

◈ Suggest a reason for not accessing psychological support.

◆ You might have suggested that obesity is particularly prevalent among women with lower socioeconomic status (Section 1.6.4) and that cost will be an issue. In addition, some people are not comfortable with accessing psychological support.

There are two forms of surgical treatment for obesity, both of which are usually referred to as **bariatric surgery**, and have the aim of reducing energy intake. The first of these is known as restrictive surgery because it restricts the size of the stomach and limits the amount of food that can be eaten during a meal. This most usually involves reducing the size of that part of the stomach that initially receives food after it is swallowed. The second approach is called reductive surgery or malabsorptive surgery because it aims to reduce the amount of digested food that is absorbed through the gut. This is achieved by greatly reducing the length of the small intestine.

◈ Why will reducing the length of the small intestine reduce the amount of digested food absorbed?

◆ Most of the digested food is absorbed in the small intestine (Section 3.3.4). If the length of the small intestine is greatly reduced, digested food may leave this section of the gut without having had the opportunity to be absorbed.

At this point, you may be thinking of another surgical procedure that is often associated with obesity. **Liposuction** is a technique in which adipose tissue is removed from specific body areas by using a surgical vacuum device. Liposuction is not generally considered by the medical profession to be a treatment for obesity. Read Vignette 9.1 (overleaf) and see whether you agree.

◈ Why is liposuction not considered a treatment for obesity?

◆ Liposuction does not treat any aspect of the conditions that cause obesity.

Liposuction is carried out for cosmetic purposes only. It makes very little difference to an individual's weight because it is not possible to withdraw much more than 1 kg of adipose tissue by this mechanism – so it cannot make an obese individual non-obese. It does not lead to altered behaviour, so it does not result in subsequent weight loss either.

◈ Why did blood ooze from areas where fat had been removed?

◆ The vacuum device sucked out adipose tissue, which comprises adipocytes together with their blood and nerve supplies (Section 1.2). This is why there was bleeding (and discomfort).

In contrast, bariatric surgery, although invasive, does not directly touch or reduce adipose tissue. So there is no immediate effect on obesity. However, as already stated, it aims to effectively alter one of the direct causes of obesity – energy intake – thus subsequently reducing adiposity.

Vignette 9.1 Donna tries liposuction

Donna, like many other undergraduates, had put on weight during her first year at university. Although Donna had only gained 3 kg in her first year and vowed to get rid of it over the summer holiday, her behaviour exemplified the weight-cycling dieter (see Box 7.1). By the time she left university, she was 5 kg heavier than when she had arrived; although she was by no means obese, she knew that she was definitely overweight and was unhappy with her size.

After completing their final examinations, Donna and her friends remained in their flat because the tenancy agreement ran for several more months. Although she had intended to return home to the north of England, Donna found a well-paid job locally and so she stayed on with her friends. Three years later she had gained a further 2 kg. Then, after splitting up with a boyfriend, she decided that she would do something positive to cheer herself up: liposuction!

From an advertisement in a fashion magazine, she found a clinic that would 'remodel' her waistline. They showed her diagrams of how removing small amounts of subcutaneous fat from the right places would dramatically alter her appearance. The cost was substantial – £3000 – but she was sure it would be worth it and decided to go ahead, despite her friends telling her she was fine as she was.

When she returned home after the operation, she was in considerable discomfort. The areas from which fat had been removed were swathed in bandages and she had been told to be sure to keep these dry. This limited her ability to wash herself and her long hair. She felt dirty and soon blood was oozing through the bandages, making her feel far worse. Healing was slow and the final result was a huge disappointment. She looked a little slimmer but not much different really and at most she was 1 kg lighter.

9.3.2 Restrictive surgery

The most common form of restrictive surgery involves placing a small inflatable band around the top of the stomach. It is known as *adjustable gastric banding*. As shown in Figure 9.10b, a small tube runs from the inflatable band to a small port (reservoir and access point) that lies just below the skin. The whole device is filled with a sterile salt solution. The advantage of this procedure is that the degree of constriction produced by the band can be altered without a further surgical procedure. Instead it is 'simply' necessary to insert a syringe through the skin and into the port and either remove some of the salt solution or add a little more. The surgical procedure itself can often be done using a *laparoscopic* technique. This involves making a small incision and then inserting a tiny camera, light source and very fine surgical instruments to carry out the actual procedure. This technique has several advantages, including more rapid recovery and a reduced likelihood of subsequent hernias – where a loop of intestine

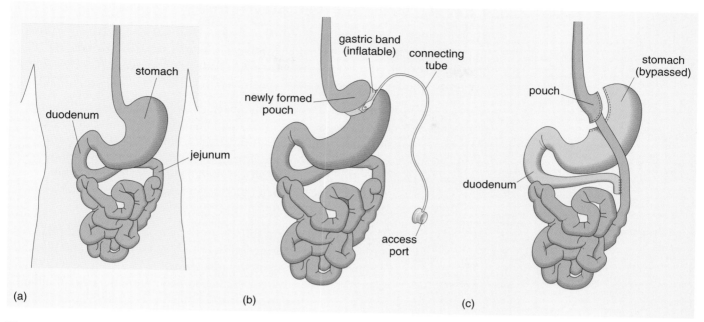

Figure 9.10 (a) The normal arrangement, route and capacity of the stomach, duodenum and jejunum. (b) Adjustable gastric banding. (c) The Roux-en-Y gastric bypass – named after the Swiss surgeon Cesar Roux and the Y-shaped arrangement of the small intestines – which combines restriction of stomach size to create a small pouch which is then connected to the end of the duodenum.

protrudes through a break in the muscular wall of the abdominal cavity. It is now widely used for a whole range of surgical procedures, not just those involving the digestive system.

An older form of restrictive surgery involved permanently restricting the size of the stomach by means of a series of surgical staples. This is a procedure that is now used much less often because it is more likely to lead to problems requiring further corrective surgery than is the case for adjustable gastric banding. However, gastric banding is not problem-free. Not infrequently, further surgery is needed if, for example, the band slips.

Restrictive surgery can be very effective in promoting weight loss. After recovery from the operation, the capacity of the stomach is typically around 100 ml, and only about 50 g of food can be eaten at one time. It needs to be moist and well-chewed, or it is liable to cause discomfort. However, it is important to be aware that, even after this type of surgery, an individual has to take a healthy attitude towards their diet. It is still possible to keep energy intake high by eating frequently and continuing to eat high-calorie foods or liquids – the latter will pass easily from the small stomach pouch and into the remainder of the digestive system. Nevertheless, clinical trials suggest that loss of as much as a half of excess weight is not unusual, and longer-term follow-up studies suggest that, although there may be some weight regain, this is much less of a problem than the regain following withdrawal of drug therapy.

9.3.3 Reductive (malabsorptive) surgery

An alternative, and more drastic, form of surgery for obesity involves bypassing part of the small intestine so that the absorption of digested food is reduced. At present, this form of surgery is usually combined with restrictive surgery using a procedure known as the Roux-en-Y gastric bypass, illustrated in Figure 9.10c.

Although the gastric bypass procedure is more complex than gastric banding, it may have some advantages. However, there are also important potential side effects, including the possibility of nutritional deficiencies as a result of the impaired absorption of food. Anaemia is another common example of this type of problem, arising through a restricted absorption of iron from the diet. One positive outcome is that patients experience a decrease in appetite.

Data from long-term follow-up studies have suggested that early weight loss is greater than for gastric banding and that this weight loss is substantially maintained over a long period. One study that came to this conclusion had enrolled a large number of obese Swedish participants. They had a mean (average) BMI of about 40 at the beginning of the study and about half received a surgical intervention. Individuals were not allocated at random to the different conditions, but instead on the basis of clinical judgement and their personal preference. The control group were reported to have received a wide range of treatments, including intensive behavioural interventions and dietary advice. Figure 9.11 shows some of the data from the 2004 report (Sjöström et al., 2004) for participants who had been followed over 10 years.

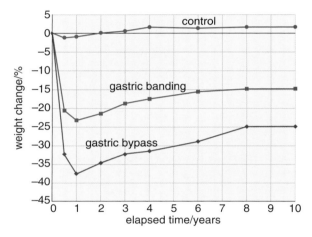

Figure 9.11 Weight change over 10 years following gastric bypass or gastric banding.

◆ What are the similarities in weight change for the different surgical groups?

◆ They show rapid weight loss during the first year, followed by a slow increase in weight during the subsequent 9 years.

◆ What is the most obvious differences between the two surgical groups?

◆ Although the broad pattern of weight loss is similar, it is clear that gastric bypass leads to a greater weight loss at all timepoints.

There were several other interesting features in the data reported in this study. At the end of the 10-year period, individuals who had had surgery were more likely than those in the control group to be engaging in active exercise during their leisure time. This might suggest that they had increased their energy expenditure. In addition, energy intake, assessed through a diet questionnaire, had decreased in the surgical groups. There was a suggestion that the decrease was greater in the group that had a gastric bypass, which would be consistent with their reduced ability to absorb food through the gut. An important additional result from this study was reported in 2007. Those in the surgical groups were less likely to have died during the study period – the probability of death was reduced to about three-quarters of that in the control group (Sjöström et al., 2007).

In recent years, there has been an increased emphasis on the overall extent to which an individual has benefited from a particular procedure. For obesity surgery, one method of evaluating this is known as BAROS and uses a scoring table to record the degree of weight loss, changes in medical conditions associated with obesity, and overall quality of life. The detailed scores are classified on a five-point scale from failure to excellent, depending on the degree of *change* from before to after the operation. Both gastric banding and gastric bypass procedures have been positively evaluated in this way. In the case of gastric banding, a Swiss study (Steffen et al., 2003) found that 89% of patients with significant associated medical conditions fell into the three upper points of the BAROS scale, with the outcome rated good, very good or excellent. The equivalent value was about 73% for individuals who were not experiencing additional medical complications associated with their obesity.

◆ Why do you suppose that outcomes for patients with complications were, in overall terms, more successful than for those without significant complications?

◆ The BAROS score depends on the extent to which quality of life *improves* as a consequence of the operation. Patients with many complications prior to their surgical treatment may have these alleviated (or not). Patients without such complications are more likely to either show no change or actually have a worse quality of life as a result of the side effects.

9.3.4   The psychological consequences of surgery

Despite the positive outcomes from surgical treatments in terms of weight loss, medical conditions and quality of life reported in the previous section, submitting to surgery is a momentous decision for each individual. Before surgery, a patient must sign to indicate their 'informed consent'. Just as with participants in experiments (Section 5.3.1), patients must have the risks of procedures explained to them prior to surgery and then sign their consent for the operation to go ahead. From a number of sources, it is clear that many patients have not fully appreciated that surgery does not provide a 'quick fix' for their problem.

Immediately after the operation, there is likely to be post-operative soreness and the patient will feel generally unwell. These factors are likely to lead to a considerable reduction in appetite and food consumption, leading to rapid initial weight loss (Figure 9.11). In addition, they are likely to be especially aware of their obesity and their decision to undergo surgery. This may help them to change their eating habits. After some months of recovery, these issues may seem less immediate, leading to a return to earlier patterns of eating.

Activity 9.1   Meal sizes after restrictive surgery

Allow 10 minutes for this activity

Use a measuring jug to help you to design three meals, each of 100 ml, suitable as a day's food for an individual to eat after restrictive surgery. What guided your choice of ingredients?

Now look at the comments on this activity at the end of this book.

◆ What will happen if an individual consumes a meal that is in excess of their stomach's new capacity?

◆ Some of the food will be forced back into the oesophagus (reflux; see Section 3.3.2).

Distressingly, almost everyone finds this happens to them initially. Often they will actually vomit after a meal, however hard they try to ensure that they are not eating too much. Having thought about the size of meals that can be eaten after restrictive surgery, you are perhaps realising that many of these patients may never be able to eat a normal meal again and that this will impinge upon their social life. Eating a meal with friends will no longer be a possibility for them, and even meeting for a coffee or going for a drink will require careful thought.

◆ In what way will meeting friends for a drink be easier than going out for a meal?

◆ Fluid will pass through the stomach more rapidly than food (Section 3.3.3). Also, meeting for only a drink, rather than for food, may avoid the need to explain the details of the surgical procedure and its effects to friends who are not already aware of them.

It is the fact that fluids pass through the stomach more rapidly than solid food that enables some people to negate the value of their operation if – or when – their appetite returns. They will liquidise food – even bars of chocolate – to enable a greater calorific intake. It may seem strange that someone would behave in this way after having voluntarily subjected themselves to bariatric surgery. In Chapter 1 it was suggested that some people eat as a response to negative events in their lives. Others may have particular types of eating disorder – binge eating was mentioned in Chapter 7. Surgery, of itself, will not reduce such people's desire to eat in response to difficult circumstances. Section 1.4.2 also raised the question of whether overeating was a response to a psychological problem or was itself a psychological problem. Interestingly, as judged by explanations in subsequent editions of a major medical textbook, the medical view has changed from seeing overeating as the primary problem (in the 1940s, the obese person was characterised as lacking in control: the slothful glutton) to seeing obesity as more frequently an individual's response to a difficult environment. Although nobody would suggest one underlying cause for all instances of obesity, a study of 340 morbidly obese people who were candidates for bariatric surgery (mean BMI of 51) found that 69% of these individuals reported that they had experienced childhood maltreatment, and of these 32% reported sexual abuse. These values are about 3 times the levels reported by non-obese groups (Grilo et al., 2005). This adds force to the guideline that surgery should not be undertaken without appropriate psychological support. If eating was a 'psychological crutch' before surgery, then the individual must have some other strategy in place for difficult life events after surgery.

9.3.5   Summary

Surgery is not recommended for individuals with BMIs of less than 35, even when they have additional medical complications that are being exacerbated by their obesity. Although there are a number of available procedures, two are widely used: gastric banding and gastric bypass. Long-term follow-up studies have shown that both procedures lead to substantial weight loss that is usually maintained over at least 10 years. The gastric bypass procedure is more effective but also more invasive and liable to side effects. In particular, there may be a substantial psychological impact of bariatric surgery that only becomes apparent in the months and years following the operation.

9.4   Summary of Chapter 9

9.1   Pharmaceutical drugs and bariatric surgery are potential clinical options for some more serious cases of obesity.

9.2   Currently licensed drug treatments rely on either restricting the digestion and absorption of fat (orlistat) or targeting certain brain neurotransmitters involved in appetite (sibutramine and, in Europe only, rimonabant).

9.3   Potential pharmaceutical targets include drugs acting at a wider range of the neurotransmitter systems involved in appetite.

9.4   Bariatric surgery may be recommended for serious cases of obesity when other techniques have failed to achieve significant weight loss and when psychological support is (or will be) available.

9.5   Weight loss following the commencement of drug treatment or after bariatric surgery is initially rapid.

9.6   Weight regain is often substantial when drug treatment is withdrawn. Even following surgery, weight regain will be observed in many cases.

Learning outcomes for Chapter 9

LO 9.1   Define and use, or recognise definitions and applications of, each of the terms printed in **bold** in the text.

LO 9.2   Describe the criteria that are usually applied to inform decisions on whether drug or surgical treatment for obesity is appropriate.

LO 9.3   Understand the rationale, in terms of supporting scientific knowledge, for a range of past, present and potential drug treatments.

LO 9.4   Describe the potential benefits and risks of surgical treatment for more serious cases of obesity.

Self-assessment questions for Chapter 9

Question 9.1 (LO 9.2)

Re-read Vignette 7.3. Would Linda meet the criteria for either drug or surgical treatment of her obesity? Give the reasons for your answer.

Question 9.2 (LOs 9.1 and 9.3)

Describe the principles underlying the use of randomised double-blind clinical trials in testing the effects of a clinical treatment. Why do regulatory bodies, such as the FDA, insist that they are used when approval for use of a novel drug is requested?

Question 9.3 (LO 9.3)

How would you describe the synaptic effect of rimonabant? Why does this drug lead to reduced body weight?

Question 9.4 (LO 9.4)

Why are the criteria for surgical treatment of obesity more restrictive than those for drug treatment?

CHAPTER 10

CHALLENGING OBESITY: THE FUTURE

If we continue to do what we have always done, then we will only get what we've always got.

French and Blair-Stevens (2005)

10.1   Introduction

For people who live in developed, industrialised societies, it is no longer 'normal' to be of healthy weight. In other words, adults of healthy weight no longer constitute the majority of the population. Worse than that, you saw in Chapter 1 how the prevalence of overweight and obesity is forecast to continue to rise and that all overweight and obese people are at greater risk of developing disabling comorbidities, hence the designation of a BMI over 25 being unhealthy. Yet the World Health Organization (WHO) points out that every country has signed up to at least one health-related human rights treaty. The WHO constitution states that 'the enjoyment of the highest attainable standard of health is one of the fundamental rights of every human being', where health is 'a state of complete physical, mental and social well being and not merely the absence of disease or infirmity'.

From this, you might conclude that within the most developed societies many people are being denied their human rights and yet are not complaining. How can we explain this? You may suggest that part of the answer to this question is that many overweight people do not consider themselves to be overweight. This is hardly surprising when such people either live in a society where to be overweight is normal or come from a culture where plumpness is prized. Even those who recognise that they are overweight or obese may not realise that they have a potential health problem. The physiological changes that disturb the body's healthy homeostatic mechanisms develop slowly and unseen, making them easy to ignore. By the time that there is no escaping the reality of a condition that has its roots in adiposity, irreversible damage may have occurred. Fortunately, however, there are a substantial number of features of some comorbidities, such as insulin resistance and hypertension, that will improve with weight loss, bringing rapid health gains.

Indeed, you might further develop your explanation of health, human rights and obesity by suggesting that it is up to the individual to take control of their own health by making changes to their lifestyle in order to achieve a healthy weight. Certainly, much of this book has considered the development and the consequences of obesity from the perspective of the individual. Thus the basic energy equation and a positive energy balance have been described in terms of the physiology and the genetics of the individual; factors influencing hunger and satiety were considered from the individual's point of view, as were the health consequences of overweight and obesity and methods used to lose weight. This focus on the individual is also seen in the media. In part, this could be because of the impact of 'human interest' stories, but it also reflects a view that obesity is an individual's problem. Eating and exercise are choices that individuals make, hence their energy balance is within their control. Or is it? Very few people choose to be fat. To be fat is to experience discrimination and stigma; children are aware of this from a very young age. But if the solution to the obesity epidemic lies with the individual to simply eat less and exercise more, why are we seeing increasing rather than decreasing numbers of obese people on the streets today (Figure 10.1)?

(a)

(b)

Figure 10.1   Street scenes in (a) 1950 and (b) 2008.

In this chapter, we will look at the underlying causes or drivers of the current obesity epidemic and will be asking what could and should be done to reverse the continuing upward trend.

10.2   The obesogenic environment

There are many who would like to blame their biology for their obesity: 'it's my glands'; 'it's my metabolism'; 'it's my genes'. You, however, know that these statements do not explain the current prevalence of overweight and obesity. For some individuals, a malfunction of one of their endocrine glands will cause metabolic disturbances that lead to obesity (Figure 10.2), but in most cases these disorders – which have a low incidence – will be diagnosed and treated successfully. Although newspapers continue to herald the discovery of 'fat genes', you know that there has not been an increase in mutations providing new variants of genes that cause obesity. Rather, there are many existing gene variants that in our evolutionary history may have aided survival but now predispose to obesity. What has changed over the last few decades is the environment, some features of which are exemplified in Vignette 10.1.

Figure 10.2   Individual with Cushing's syndrome – a consequence of prolonged excessive secretion of the hormone cortisol. Cortisol is secreted from the adrenal gland which lies just above the kidney. Cushing's syndrome may be caused by a tumour affecting the adrenal gland or the part of the brain that regulates the adrenal gland. It can also be caused by the prolonged use of corticosteroids for treatment of arthritis or asthma. Incidence of all types of Cushing's syndrome is about 1 in 5000.

Vignette 10.1   Shopping habits

Linda has just had some shopping delivered by her niece Sharon. Sharon lives on a new estate with no local shops and does her grocery shopping each fortnight at a hypermarket about 3 miles (5 km) away down the dual carriageway. She travels there by car and parks within 50 m of the entrance. The shop is heated and contains a café. So having chosen her items and placed them in the trolley, she usually has a mid-morning coffee and scone. Today, however, tempted by Auntie Linda's delicious selection of cakes, Sharon visits her for coffee and cake. Sharon then drives back to her warm house, puts the shopping in the freezer, fridge and larder, loads and turns on the washing machine, then settles down to watch television.

For some reason, seeing Sharon today has reminded Linda of Gladys, her grandmother, and she remembers how Gladys had to walk to the shops. In the 1960s, Gladys did her shopping locally. There was a small parade of shops about a quarter of a mile (400 m) from her house and she shopped at least three times a week to maintain a supply of meat, vegetables and bread; she had a small fridge, but no freezer. The shops were small and, apart from the bakers, not much warmer than the temperature outside. There was a tea shop, but she visited it for special occasions once a month when she socialised with Mary and Edith. Often Gladys would catch the bus for the short trip back, especially when she bought potatoes or other heavy produce. Her house was chilly – no point keeping the gas fire lit if she was going out shopping – and anyway polishing and scrubbing kept her warm. But after shopping she would usually make a cup of tea and turn on the fire in the living room while she sat down for a bit and got on with her knitting. Linda thinks that Sharon looks a lot like Gladys except that she hasn't inherited Gladys's slim figure.

◆ Consider the differences in the way the two women in Vignette 10.1 shop for the everyday necessities and how their shopping behaviour affects their energy balance. To what extent is their behaviour a personal choice?

◆ Every 2 weeks, Gladys will make at least six 400 m trips – and more if she walks back – so she will walk more than 2400 m, whereas Sharon walks 50 m to reach the shops. Gladys goes to the tea shop once a month; Sharon goes every 2 weeks. It is not possible to say whether Gladys chose to walk rather than take the bus to the shops, but it sounds as though this, as well as infrequent visits to the tea shop, might be a financial necessity rather than a personal choice. It would seem that Sharon has no shops within reasonable walking distance. The provision of shopping trolleys means that Sharon does not have to consider the weight of her purchases and owning a car and a freezer means that it is practical to shop once a fortnight. In making that choice, she may limit the amount of fresh fruit and vegetables that she eats and may buy ready-meals which are often an energy-dense alternative. However, she may 'choose' to shop infrequently because she has a job and little spare time.

To a considerable extent, it may be that the differences between the ways in which Sharon and Gladys shop is not so much personal choice but a consequence of their social environment. Since the 1960s, there have been many social changes in the UK, mirrored in many other developed societies, that reduce the need for physical activity in day-to-day living and that encourage overconsumption.

◆ List as many of these changes as you can, together with their consequences.

◆ Table 10.1 lists some changes you might have suggested.

Table 10.1 Examples of social changes that have happened in the UK since the 1960s and which have resulted in overconsumption and reduced levels of physical activity.

Social change	Consequence
Increased car ownership (doubled over 30 years)	Used for commuting, school run, etc., so people walk less; roads more dangerous, so children cycle less and do not play in the streets
Media-driven perception that children are at high risk of abduction and attack	Children play less actively because they are kept at home and indoors
Increased ownership of appliances such as computers and televisions (viewing has doubled over 30 years)	Leisure activities, particularly for children, are more sedentary
More televisual advertising of energy-dense food – particularly at times when children watch	More energy-dense food eaten
More sophisticated machinery	Fewer strenuous manual jobs with high energy intake required
More working mothers	Mothers have less time to prepare meals, and instead buy more ready-meals and snacks that tend to be energy-dense
Increased availability/affordability of washing machines, vacuum cleaners, etc.	Housework is easier
More meals purchased outside the home, e.g. ready-meals, takeaways and restaurant meals	Increased consumption of generally energy-dense food
More variety of foods and flavours available	People eat more
Portion sizes have increased	People eat more
Housing estates built without local facilities such as shops, parks and playgrounds	People drive to shops, parks and playgrounds
More out-of-town shopping centres	People drive to shops
Disability legislation means more public buildings have lifts	Able-bodied people also use the lifts
Houses, shops and places of work are warmer	Do not need to keep active to stay warm, less energy required to maintain body temperature

So far in this course, we have given evidence of some of these changes and the way they may affect an individual's chances of becoming obese. Evidence that the other factors may be important can be found in a number of government-commissioned documents that seek to understand the causes of obesity and the ways in which individuals are at the mercy of an environment which they did not create. In the next section, we look at some of the reasons why governments are trying to understand the causes of the obesity epidemic.

10.3 Obesity and government

The point was made in Section 1.5 that obesity incurs costs to public health services but that these are not the only costs to the public purse.

◆ Study Table 1.4; are health care costs the major portion of the overall estimated financial cost of overweight and obesity to the UK Government?

◆ No, the indirect costs attributable to sickness and mortality are twice as great as the direct costs to the NHS.

The costs attributable to sickness represent the loss of tax receipts from the lost earnings of individuals and sick pay as a consequence of certified sickness. This is probably a substantial underestimate because it does not include uncertified sickness. Nor does it take account of the fact that the chance of being employed are lower – a reduction of maybe as much as 25% – for someone who is obese when compared with someone of healthy weight. This is part of the price of discrimination, as discussed in Section 7.3.1. Other consequences of these lower levels of employment and long-term sickness are the need for government agencies to make incapacity or unemployment benefit payments and to provide social care – for example, home help. Again, no proper estimates are available for these costs, but suggestions are that, in total, the indirect costs of overweight and obesity are at least double the £5 billion shown in Table 1.4. As the prevalence of obesity escalates, so do estimates of the proportion of the national budget that will need to be committed to dealing with the consequences.

Cost alone does not require the government to seek to understand the causes of events, nor does it justify framing policy to change events. However, there are other grounds that should be examined, with suggestions such as the upholding of laws in relation to discrimination and the protection of vulnerable individuals. If a child is obese at 6 years old, their chance of being an obese adult is greater than 50%; habits relating to food eaten and exercise taken are established early in life and are subsequently hard to change. Given these facts, it might seem reasonable to expect government intervention to combat childhood obesity. However, other arguments for intervention can be brought forward, but even in the case of the two minimally contentious suggestions just given, there will not be universal agreement on the government's rights to intervene, nor what form that intervention might take. The next section takes a look at some of the issues that are relevant to framing and evaluating social policy.

10.4 Evaluating policies to reduce the incidence of obesity

10.4.1 Identifying possible policies

In this section of the chapter, we will look at some of the issues that arise when trying to decide what might, and might not, be the appropriate policies to help reduce current levels of obesity in the UK and other countries. We will approach the problem in two different ways. The first of these is to see whether **ethics**, the branch of philosophy which develops frameworks to decide what is right or wrong, might be of some help. For example, could an ethical framework help in deciding whether it would be appropriate to refuse medical treatment to obese people, to tax junk food in the same way as cigarettes, or simply leave individuals to make their own choice in relation to the foods that they eat and the exercise that they take? The second approach may seem much more down to earth. It simply involves asking a wide range of **stakeholders** what they would regard as appropriate policies. In this context, the term stakeholder refers to representatives of a whole range of organisations within society who might be expected to have views on the topic. This includes representatives from government (especially from finance and health ministries), from pressure groups with a relevant interest (perhaps from farming and health-related areas) as well as from industrial and other commercial concerns (for example, food manufacturers and retailers). If a very structured approach is taken to interviewing these different individuals, then it is possible to extract data which show the extent to which particular policies might be broadly acceptable. However, before looking at these two approaches in more detail, explore your own attitudes to possible policy options in the following activity.

Activity 10.1   Ideas to halt the obesity epidemic

Allow 30–60 minutes for this activity

List a series of policies that government or non-governmental groups (including a range from pressure groups to food retailers) could implement which might have a beneficial impact in relation to current obesity levels. Try to be as imaginative as possible, including in your list policies of which you personally would disapprove. For each policy, write a short sentence saying whether or not you approve of it, and whether or not you think it would be effective. Later in this chapter, you will use this list to evaluate the two approaches to policy development that we have outlined in the introductory paragraph to this section.

Now look at the comments on this activity at the end of this book.

10.4.2   Evaluating policies to combat obesity: an ethical approach

Policies that might help to reduce the increased rates of obesity that have been suggested in the past few decades are numerous and diverse. There are at least two important questions that should be addressed for any particular policy. One question is whether it is ethical, or right, and the second question is whether it is likely to work. This section examines the first of these questions.

Any consideration of the ethics of public health policies has first to agree a general framework on how to decide what is right or wrong. There is, of course, an enormous variety of approaches, but it may help to consider some more extreme positions before outlining a framework that it is likely to have general acceptance. It is possible to imagine a society which values individual freedom above anything else. In such a society, individuals might be allowed to do anything they wish, providing it did not actively harm another individual. A society of this kind would be described as **libertarian**. Clearly, in a society that values individual liberty above any other consideration, taxing junk food and a whole host of other policies would be seen as inappropriate because they interfere with such liberty. Indeed, educational programmes, paid for out of taxation, would also be likely to be ruled out. In such a society, why should an individual be exposed to the guilt that might arise from such programmes? Why should they be expected to pay for it? However, a society of this kind also has to face some very difficult choices. What action, if any, should it take to help those who have exercised their freedom in a way that has led to ill health? In fact, a libertarian society of this kind would not provide state-funded health care or education. That would be a matter for the individuals living in that society. For this kind of reason, extreme libertarian models are seen to be very problematic for a society which cares about the welfare of all of its members.

It is possible to imagine a society where decisions are made on a very different basis. Suppose that for any given policy it were possible to work out (at least roughly) the benefits and the costs to that society. The imposition of a tax on junk food would deprive (or, at least, restrict) some individuals' ability to eat whatever they choose. On the other hand, assuming the society has a state-funded health care system, it will reduce the cost of caring for individuals who have developed obesity-related diseases as a consequence of eating junk food. The principle of determining the costs and benefits of a particular policy and then using this information to decide whether or not the policy is acceptable is sometimes described as **utilitarian**. These ideas are associated with the English philosopher Jeremy Bentham (1748–1832) and were embraced, developed and brought to public attention by his pupil John Stuart Mill (1806–1873).

◆ Taking the example of taxing junk food, what problems can you foresee in trying to calculate the costs and benefits in order to decide whether such a tax should be imposed?

◆ One obvious problem is that any calculation requires that the different components (here, the costs and benefits) should be measurable and expressed in the same units. For example, it may be possible to estimate health care costs under different options, but if one option involves restrictions on personal freedom, then it may be hard to give that a monetary value.

Activity 10.2 Libertarian and utilitarian policies

Allow 10 minutes for this activity

At this point, go back to your list of policies from Activity 10.1. For each one, decide whether you think it would be acceptable under a (i) libertarian or (ii) utilitarian framework, giving a simple 'yes' or 'no' answer.

Now look at the comments on this activity at the end of this book.

At this point, you may be wondering whether there is some more reasonable halfway ground between these two positions. In fact, it is frequently argued that liberal democracies of the kind that are found in Europe, North America and in many other states throughout the world do occupy this middle ground. Key principles of a liberal state include:

- concern for, and value of, individual freedoms
- concern for the welfare of all members of the society.

A part of the concern for general welfare relates to health and, although these two principles may seem wholly acceptable, it is not hard to imagine health-related examples where they contradict. Take, for example, the ban on smoking in public places that was introduced in the UK during 2006–7.

◆ How does this ban on smoking arise from one of these principles but contradict the other?

◆ It is consistent with the second principle (concern for welfare) because passive smoking is now known to have significant health risks associated with it, but it takes away from the freedom of individuals who wish to smoke in this way.

So one important question in relation to a society's concern for welfare must be how far it can go when the policies that are being considered will also restrict individual freedom. One answer to this question, which is still very influential, was provided by John Stuart Mill in his essay entitled *On Liberty* (Mill, 1859). In simple terms, he suggested that restrictions on personal freedoms can be justified when to impose them will avoid harm, or ill health, occurring to another individual, but are not justified when their imposition would simply enhance the health of that individual. This is often referred to as Mill's **harm principle**. Mill also suggested that additional safeguards were necessary in relation to children, or more vulnerable adults. He felt that education in relation to possible harms was appropriate but that this should not extend to attempts to coerce individuals to behave in a particular way. In 2007, a publication from the Nuffield Council on Bioethics (NCB) took these ideas and extended them to produce an ethical framework for public health that is referred to as the **stewardship model**. The model gets its name from the idea that the role of states, their governments and even (at least to some extent) non-governmental bodies such as companies is to act as stewards to the needs of people. Some of these needs will be individual, perhaps relating to particular health problems, and will vary with age, gender, ethnicity, and so on. Other needs are more general, such as the need for clean water supplies, sewage disposal, and so on. The NCB report authors suggest that a public health programme should meet one or more of the following criteria:

1   aim to reduce the risks of ill health that people might impose on each other

2   aim to reduce the causes of ill health by regulations that ensure environmental conditions that sustain good health, such as the provision of clean air and water, safe food and decent housing

3   pay special attention to the health of children and other vulnerable people

4   promote health not only by providing information and advice, but also with programmes to help people to overcome addictions and other unhealthy behaviours

5   aim to ensure that it is easy for people to lead a healthy life, for example by providing convenient and safe opportunities for exercise

6   ensure that people have appropriate access to medical services

7   aim to reduce unfair health inequalities

but should accept the following constraints:

8   not attempt to coerce adults to lead healthy lives

9   minimise interventions that are introduced without either the individual consent of those affected or … (other safeguards, such as democratic decision-making procedures)

10   seek to minimise interventions that are perceived as unduly intrusive and in conflict with important personal values.

◆ Think about how this framework could be applied to some of the specific policies that you may have thought about at the beginning of this section. In particular, how would you relate the proposal to provide safe cycling and walking routes to school in terms of these proposals? For each criterion and constraint, give a rating of inconsistent, consistent or irrelevant.

◆ Such a policy would be consistent with criteria 1–5, whereas criteria 6 and 7 are probably irrelevant. The proposal also does not seem to conflict with constraints 8–10.

In this case, it seems that the policy is relatively uncontroversial, at least in terms of the ethical framework that has been outlined here. Now let's examine a proposal that may be more problematic.

◆ How would you rate the proposal to tax junk food in relation to the NCB criteria and constraints listed above? For each criterion, provide a rating of inconsistent, consistent or irrelevant. Assume that the consequence of such a tax would be to reduce consumption. (Note: this assumption will be examined in a later section.)

◆ Such a policy would be consistent with criteria 1–4 and 7, whereas criteria 5 and 6 are probably irrelevant. The proposal is in conflict with constraints 8–10.

As you will appreciate from looking at this example, some possible policies may be consistent with some criteria, but inconsistent with others. The introduction of such policies is therefore more likely to be controversial and lead to public opposition. Nevertheless, an ethical perspective does provide one way of thinking about public health policies and the extent to which they might be accepted within society and implemented by both government and non-governmental organisations.

10.4.3 Evaluating policies to combat obesity: views from stakeholders

An alternative way of developing policies to tackle the challenge of increasing obesity levels is to seek the opinions of a wide group of individuals and organisations concerned with the issue: the opinions of *stakeholders*. In this section, we shall look at the methods and results from one such large-scale consultation known as the PorGrow project (policy options for responding to the growing challenge of obesity; Millstone and Lobstein, 2007).

The study consisted of four phases:

1   choose the set of options about which stakeholders would be consulted

2   develop criteria which would be used to record the responses of these stakeholders

3   carry out the individual consultation interviews with these stakeholders

4   analyse and present the results.

This process is shown diagrammatically in Figure 10.3.

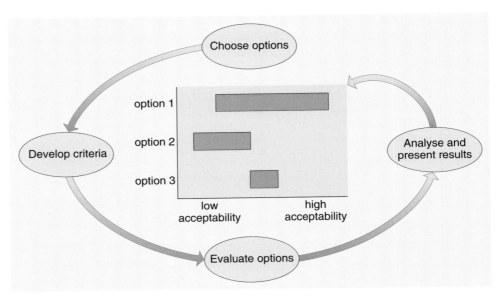

Figure 10.3   How the PorGrow project was conducted.

Early in the design of the study, which involved consultations in each of the member countries of the European Union, a list of stakeholders was identified that fell into seven broad categories. They included representatives of government and non-governmental groups, a variety of commercial groups, including large food retailers, other providers of both food and exercise facilities, as well as specialists in the area of public health. It was agreed that they would be questioned, in depth, about their attitudes towards a series a different options that might be used to combat rising obesity levels. They included changes in planning and transport policies and in the provision of exercise facilities, which might address the issue of declining activity levels. They were also consulted on a wide variety of possible measures in relation to food policy, including food labelling, the imposition of taxes on 'unhealthy' foods and subsidies on 'healthy' foods. In addition, there were questions relating to more specific issues such as increasing budgets for obesity research, the use of fat and sugar substitutes and the likely usefulness of pharmaceutical treatments of obesity. During the interviews, respondents were able to define the criteria that they would use to judge the acceptability and effectiveness of the different options. There was a small set of options that all respondents had to judge (the core options), but they had some freedom to choose to judge some of a list of additional options (discretionary options).

Following completion of these interviews, the results were analysed in a way that allowed the outcomes to be compared between different groups of stakeholders and different countries. An overall summary is shown in Figure 10.4.

◆ What option was found to be the most acceptable to the range of respondents who were interviewed for this study?

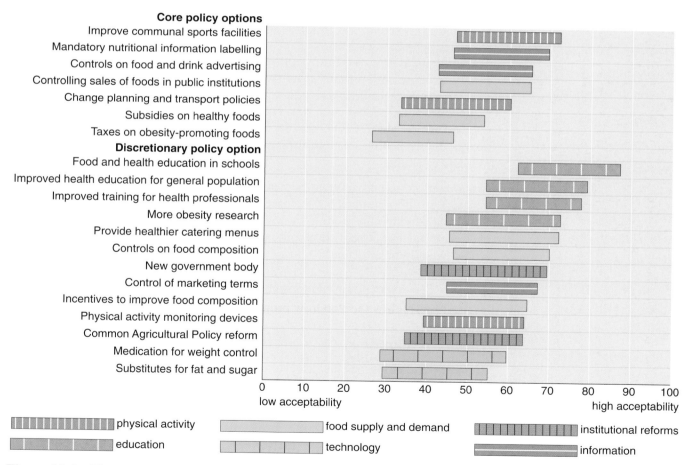

Figure 10.4 The range of responses given to a selection of core and discretionary policy options. The length of each bar represents the range of scores given to that option, from 'low acceptability' to 'high acceptability'. Interviewees had to rate the acceptability of each option under two scenarios: an optimistic one and a pessimistic one.

◆ Food and health education in schools.

◇ What option had least support from those questioned in this survey?

◆ Taxing unhealthy food.

It is interesting that, seen from the perspective of the ethical framework discussed in the previous section, the taxing of unhealthy or junk food was particularly problematic because of the way in which it might be seen to infringe important personal freedoms. Provision of education – particularly to children, who are seen as an especially vulnerable group – is consistent with all of the criteria and breaks none of the constraints of that ethical framework.

There were a great number of more detailed outcomes from the PorGrow study; as would be expected, there were significant differences between respondents in relation to some of the policy options. Pharmaceutical treatments were rated much more highly by representatives of that industry than by any others. Respondents from major food retailers were less enthusiastic than others about policies that might restrict their commercial freedom. These included food labelling and advertising controls. In addition, there was interesting variation in which options were favoured from one country to another. For example, policies that centred on activity and exercise were more favoured in Eastern European countries, whereas controls on advertising and food labelling received greater support in many Western European countries. Perhaps this is not a surprising finding because until recently there has been little advertising of food on television in Eastern European countries. There were also interesting variations at the level of individual countries. For example, it was reported that French respondents felt that 'traffic light' labelling of food was an overly simple 'Anglo-Saxon' device that was incompatible with French attitudes towards food and the pleasure of eating!

In summary, the approach represented by the PorGrow project can provide very useful information about the acceptability of different policy options and the extent to which these could be implemented across national boundaries. However, it does not provide empirical evidence as to whether particular policies would actually be effective.

10.5 Combating obesity: what policies would be effective?

10.5.1 Evaluating policy effectiveness

So far, in evaluating whether a treatment is effective in combating obesity, we have used the well-established scientific methods of experimentation – for example, examining the outcome of randomised controlled trials.

◆ Look at the policy options given in Figure 10.4 and identify any that would be amenable to having their outcomes assessed by randomised controlled trials.

◆ Although some of the options would be amenable to assessment by randomised controlled trials, the control of the trials might not be very effective. For example, if you improved the food and health education in one school in the neighbourhood, it is possible that children at that school would bring home ideas that are then shared with siblings attending a different school, or that parents talk about the scheme to other parents who have children at different schools.

The randomised controlled trials that have been reported are from settings such as schools, health centres or workplaces (microenvironments). Often expensive to run, these trials suffer from lack of long-term follow-up and national transferability. No trials have yet delivered major benefits in terms of preventing overweight and obesity.

An alternative approach for assessing whether a policy might be successful would be to use epidemiological studies that compared obesity prevalence in countries that differed with respect to the policy option that interested you – for example, obesity prevalence in two countries that had markedly different levels of communal sports.

◆ What would be the difficulty with an approach of this type?

◆ The two countries would undoubtedly differ in other major ways, for example advertising of food products on television (see previous section).

A further problem is that we are seeking to ascertain whether an *intervention* could halt or reverse the upward trend of obesity prevalence. What we need to find is a country or a region where such a change has occurred. If you look at Figure 10.5, you can see that for a period between the early 1980s and 1990s the prevalence of obesity in Cuba approximately halved. Unfortunately, this was not a result of any policy change to combat obesity, but was a consequence of an international situation that led to an overall drop in the standard of living in Cuba – to a level that no government would willingly impose upon its citizens! The situation in Finland is different; here it is possible that the halt in the trend of ever-increasing obesity prevalence is the result of a project that was set up in the 1970s to combat coronary heart disease. Apart from the fact that this was not set up as a controlled experiment, you can see that there is a considerable time lag between the start of the project and changes in the rate of increase of obesity prevalence. This is another problem relating to the assessment of the likely efficacy of any policy change: for most of the population, adiposity changes occur

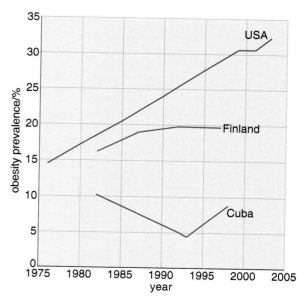

Figure 10.5   Trends in adult obesity prevalence in Finland, Cuba and the USA.

gradually, but no government wants to spend 10 or more years doing nothing while waiting to find out whether a particular policy change might be worth implementing.

Another way of assessing the efficacy of a policy option is to use a model to assist with the conceptualisation of the problem and/or to predict outcomes. Some observers suggest that, having termed the situation an obesity epidemic, it is instructive to examine the methods used to manage epidemics. An epidemic is traditionally the term for the explosive spread of an infectious disease. Management of an epidemic considers three points of an epidemiological triad (Figure 10.6): the host (person or animal who has a disease, such as malaria); the vector (the distributor of the disease or agent – often an insect such as a mosquito); and the environment (the situation where the vector can flourish – such as stagnant water).

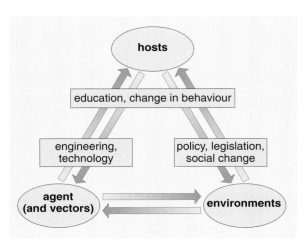

Figure 10.6 The epidemiological triad showing the relationships between the points of the triad and the types of initiatives typically used for dealing with each point.

So, in the case of malaria, resource could go to the individual suffering from the disease to alleviate the symptoms, or it could be directed towards the destruction of the mosquito which is carrying the agent of disease (a microbe), or to the disruption of the waters where the mosquito breeds – the environment. Using this model has proved useful in tackling 'epidemics' as varied as smoking and road traffic accidents (RTAs). With RTAs, for example, using this model moved the focus from the behaviour of the driver (the host) to the agent (speed) and the vector (the vehicle). Fitting vehicles with airbags, crumple zones and seat belts has had the beneficial effect of reducing the number of deaths and injuries incurred by drivers and passengers in RTAs. In the case of obesity, attention has so far been mainly focused on the host: treating individuals and trying to change their behaviour. Attention is increasingly turning to the obesogenic environment, where food is readily available and technological advances have reduced the need for physical activity. The third point on the triad, the vector and agent, is much less obvious, but the benefit of using the epidemiological triad is that it forces us to identify and deal with vectors. Obesity occurs when an individual has a chronic positive energy balance, so this chronic positive energy balance is the agent. The vectors, therefore, are related to both sides of the energy balance equation. So energy-dense foods, sugary drinks and large portion sizes are the vectors for energy intake, and for energy expenditure the vectors are the many labour-saving and time-saving devices, together with the passive entertainment brought by modern technology.

For other epidemics, vector management has involved technologies such as the development and spraying of insecticides to kill mosquitoes and the modifications to vehicles described above to minimise the impact of speed in traffic injuries. In the case of obesity, although technology has caused many of the problems, it can nevertheless be harnessed to challenge the development of obesity at population levels. Food technology aimed at enhancing the flavour of low-energy options is one obvious idea. Another very practical suggestion comes from New Zealand, where it was noted that the average fat content of deep-fried chipped potatoes sold at takeaways was 11.5%. This is higher than it needs to be; using optimal cooking techniques, the fat content could be reduced to 10%. By training staff in these techniques, it is calculated that a very considerable benefit could be achieved because the mean annual consumption of takeaway chips is about 40 kg per person. Thus a 1.5% reduction in fat content would result annually in an estimated decrease in fat consumption of about 0.5 kg per person. Currently the average annual weight gain of adults in New Zealand is around 0.3 kg so, in theory, this measure alone could reverse the trend of rising obesity prevalence in New Zealand.

As with manipulation of vectors, acting to alter the environment has its effect at the population level rather than the individual level. Environments are very complex systems and are often difficult to conceptualise. One device for analysing obesogenic environments is the ANGELO framework (analysis grid for environments linked to obesity; Table 10.2). This enables policy makers to keep the 'big picture' to the forefront of their thinking while assessing possibilities in relation to one part of the environment. The grid shows that there are a number of types of environment that need to be considered when engaging with the possibility of assessing evidence and planning interventions at, say, the microenvironmental level – such as within a school or a defined neighbourhood. For example, the environment could be physically altered by removing vending machines, but if parents believe that their children should have access to such snacks and the school rules allow children to leave the premises to purchase snacks locally, then the policy will be ineffective. Furthermore, the school may bear a financial penalty by losing revenue from housing the vending machine.

Table 10.2 The ANGELO framework which is used to encourage holistic thinking about the obesogenic environment when planning new policies to combat obesity.

| Environment type | Environment size | | | |
| | Microenvironment (settings) | | Macroenvironment (sectors) | |
	Food	Physical activity	Food	Physical activity
Physical What is available?				
Economic What are the financial factors?				
Policy What are the rules?				
Sociocultural What are the attitudes, beliefs, perceptions, values?				

While using the ANGELO framework to work through the set of questions to establish how a particular policy might affect the different aspects of food supply in a setting such as a school, there is a constant reminder from the right-hand side of the grid of the macroenvironment (sector) within which the setting operates. The macroenvironment must also be changed – indeed, changing the macroenvironment could render as superfluous some policies for the microenvironment. For example, if energy-dense snacks were withdrawn from manufacture by the food industry and instead they made high-fibre, low-energy snacks desirable via advertising, it might not be necessary to remove the school vending machine.

Thus the epidemiological triad used with the ANGELO framework encourages the taking of comprehensive action to combat obesity. Nevertheless, modelling small parts of the environment can also be of value, as is shown in Figure 10.7, taken from a Danish study.

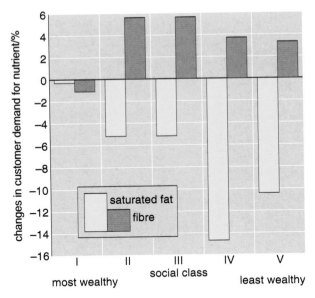

Figure 10.7 Manipulating sales taxes in Denmark could benefit the diet of the least wealthy: the likely changes in consumption of fibre and saturated fat that would follow tax changes. These changes involve a reduction in tax from 25% to 22% on fruit, vegetables and wholegrain; an increase in tax from 25% to 31% on butter, cheese, pork, beef and fatty meats; and a new tax on sugar.

◆ Look at Figure 10.7. Which social class would have the greatest decrease in intake of saturated fat?

◆ Social Class IV (one of the less wealthy groups) shows the biggest percentage decrease (15%) of saturated fat intake.

Although it was suggested earlier that taxing junk food was problematic, note that the Danish study is modelling a fairly small shift in taxation. These suggestions could be viewed as merely redressing the balance in the context of overall increases in the cost of the healthy foods versus the obesogenic components of our diets (Figure 10.8).

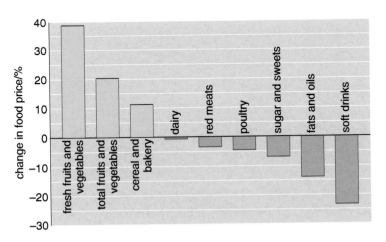

Figure 10.8 Changes in some food prices over a 15-year period (1985–2000) in the USA.

10.5.2   What policies have been tried?

Most resources have, so far, been aimed at changing individual behaviour. The statements made by the US politician Tommy Thompson that US citizens were 'too darned fat' – quoted at the beginning of this book – were made in the context of announcing that the US government was planning to put more money into research and into public education. In particular, in response to the perception that most of the 64% of US citizens who were overweight felt their situation was insuperable because only radical lifestyle changes could alter their size, a 'small steps' campaign was initiated. This encouraged individuals to make small, sustainable changes to their everyday lives, such as using stairs rather than escalators and snacking on a piece of fruit rather than a packet of crisps. In the UK, an NHS booklet published in 2006, *Your weight, your health*, echoed the message with a section entitled 'Small steps – where to start'. In fact, statistics suggest that making small changes can be very beneficial because a sustained weight loss of 5–10% makes a worthwhile improvement to health risks. You have read that at the population level such initiatives have yet to make an impact and that at an individual level many of those who lose weight subsequently regain all, or more, than they originally lost; despite this, there are some individuals who do successfully lose weight and maintain the loss. Typically, the successful dieter is one who does take small steps, setting manageable, long-term targets, thinking in terms of years rather than weeks (see Vignette 10.2). These people usually have good **social capital** – a term used to describe how effective support networks provide benefit to the individual. For example, in Chapter 8 you saw that benefit came from joining groups such as WeightWatchers and from working with the family group, not just with one child.

Vignette 10.2 Linda enjoys Florida

Holidaying in Florida, Linda noticed many other people of her size who were obviously enjoying life. She began to feel better about herself so when, the day after her return, an old friend phoned and asked if Linda could do some voluntary office work for an animal sanctuary, she agreed. Linda also decided to visit the doctor again about her arthritis, which was much less troublesome while she was in Florida. A new doctor sees her and suggests she joins a weight management group that will be starting the following month. The doctor tells Linda that she should think in terms of losing weight over the next 2 years and sets her the goal of losing about 0.5 kg a week. The doctor also arranges for Linda's painful joints to be X-rayed and recommends some non-prescription pain relief tablets.

Find out how Linda got on by answering Question 10.1 at the end of the chapter.

At a practical level, it can be difficult to change individual behaviour. Much behaviour is governed by habit rather than by active consideration of choices that could be made. Even when the individual wants to change, social norms may constrain their response. Sharon expects Linda to offer cake (Vignette 10.1), and in the office you are expected to accept a piece of birthday cake (Section 8.6).

Education is one way to bring about change; again, this is aimed at the individual, but often advice on healthy living is, apparently, ignored. Research into why guidance on health issues is not followed identified that messages are often not reaching the targeted audience, who therefore remain in ignorance. In Chapter 8,

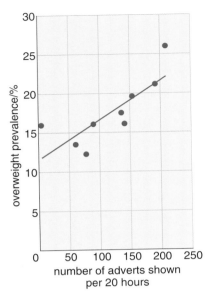

Figure 10.9 Correlation between overweight in children and frequency of advertising for obesogenic items in seven EU countries, Australia and the USA.

you learnt how easily wrong messages could be picked up (Figure 8.10) and there is a further issue around the conflicting messages within the environment (compare Vignettes 1.4 and 10.3). For example, the majority of messages about food are promotional advertising, not healthy eating advice. Spending on advertising by the food industry in the UK is 100 times greater (at £743 million) than the amount spent by government on health education (£7 million), and 70% of the food advertisements on children's television are for obesogenic items. This may have an effect on children's behaviour (Figure 10.9), and some countries currently (2008) have controls on advertising on television aimed at children (Sweden, Greece and Ireland); in France, there is a tax on all soft drinks advertising; and Brazil allows no advertising for unhealthy foods in the vicinity of schools.

Vignette 10.3   Maureen supports Charlie's new diet

Maureen and Dave are delighted to learn that Charlie is in the school swimming team. Maureen looks at the healthy eating sheets supplied by Mr Boyce and realises that she needs to make some changes to her shopping list – no more crisps and treacle tart! Now that Charlie is training after school most days, she is often home before him and has devised meals that she can put on the table within minutes of his arriving home so he doesn't need a snack. The healthier diet has benefited the whole family – both she and Dave find they have more energy these days.

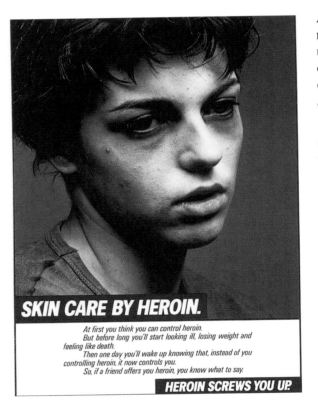

Additionally, for healthy advice to change behaviour it is necessary for the individual to feel that the required changes are under their control. So a mother might feel it is pointless to buy expensive pieces of fruit with a relatively short shelf life if her children cannot be persuaded to eat them.

The government cannot simply give advice; it must also 'market' the idea. **Social marketing** – whereby promotion of the acceptability of an idea precedes behavioural change – has been used successfully in other government health campaigns such as smoking and drink-driving. One such local initiative aimed at challenging childhood obesity is 'Water is cool in school', a project against soft drinks sales – as yet (2008) unevaluated. The language and images required for the successful marketing of an idea to children and young people can be very different from those used for marketing to adults and needs careful research and implementation, as Figure 10.10 demonstrates.

Figure 10.10 A 1986 advertisement used by the UK's Health Education Authority to warn of the dangers of heroin. Unfortunately, the image was used as a 'pin-up' by teenage girls, who found the look 'sexy'.

Although research has not been well funded, the challenge of the obesity epidemic is forcing governments to push obesity issues higher up the agenda. One outcome of the UK Government-sponsored 'Tackling Obesities' project was the 'Obesity System Map' (Figure 10.11). This map shows the links between a multitude of factors that together influence an individual's energy balance, and clearly demonstrates that the issue cannot be tackled by any one government department, instead requiring cooperation and coordination among government departments, industry and voluntary organisations. Many organisations are already thinking seriously about their roles, an example being the Institute of Architects (USA), who held a conference on their contribution to reducing the obesogenic environment.

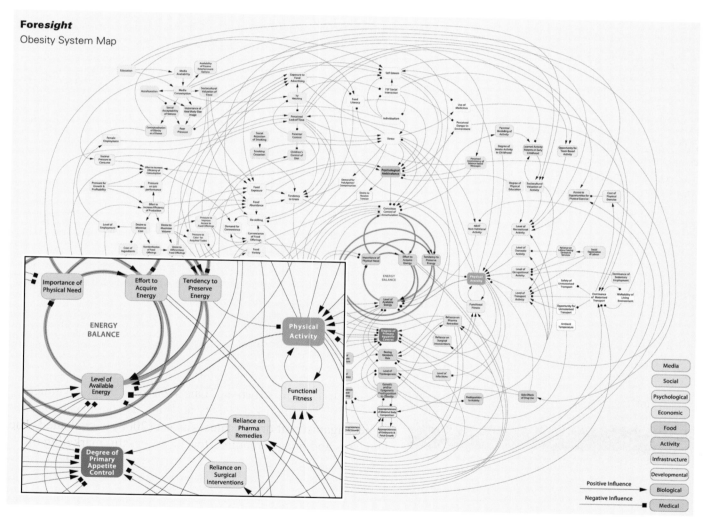

Figure 10.11 The Obesity System Map, with an area shown in detail in the inset.

10.6 Conclusion

Obesity is not an inevitable consequence of 21st-century living; rather, it is a consequence of choices made at many different levels and must be challenged on multiple fronts. There needs to be consideration given to those who are already overweight or obese at the same time as working towards preventing others from moving into that category. Childhood obesity needs to be seen as a separate issue, because childhood obesity predicts both adult obesity and an early onset of comorbidities and because the rate of increase in prevalence is itself increasing.

Individuals can and should take steps to lose weight and increase activity, but this is hard when advice and support are not always forthcoming. Even health professionals working in areas related to obesity can hold discriminatory views on obesity, and communication failures can pass unnoticed. However, many medical schools have changed their training and have demonstrated that this improves their student's attitudes towards obese patients. Support is needed from the wider population too. Overtly abusive behaviour towards obese people is relatively rare, but discrimination and social exclusion are widespread. The claim that this is 'motivating' and will encourage weight loss is untrue. As you have read, positive plans and encouragement are far more likely to lead to permanent changes in behaviour. It is also the case that learning about the biological basis of the development of obesity can create a more helpful and sympathetic public. Maybe this course has challenged and changed your own attitudes; writing it has changed ours.

It is hard to escape the fact that, although there has been a tremendous increase in understanding of the biological systems and processes that underlie obesity, this has not led to any check on the obesity epidemic. It is estimated that at any one moment around 70% of the adult population in the UK is trying to lose weight or, at least, not put on any more weight. Clearly, this is not getting us anywhere in the battle against obesity. Humans evolved in conditions where food supplies were uncertain and meagre and are best adapted for living in such conditions. The modern obesogenic environment and the behavioural response to it are the causes of the obesity epidemic, and so eradicating the epidemic requires a better understanding of how behaviour and/or environment might be changed. There can be no expectation of any simple solutions – other than the horrors of famine – or of any 'quick fixes'. In planning any policy change, it is necessary to have an awareness of the differences between individuals in their biology and their social and cultural attitudes towards obesity and the obesogenic environment. An integrated approach to the issues raised by obesity is likely to require a range of policies which need to be initiated at the level of government but which involve many other national organisations. These are just some of the factors that are involved in challenging obesity.

10.7   Summary of Chapter 10

10.1   The obesogenic environment encourages a positive energy balance through the adoption of behaviours over which individuals have little control.

10.2   The obesity epidemic carries a high cost for governments.

10.3 There are many possible policy options to combat obesity. Their acceptablity can be assessed using an ethical framework and involving stakeholders in decision making.

10.4 Initiatives that might inform policy have usually not been evaluated over a sufficient time period and most have had little effect.

10.5 Modelling is an alternative approach that has been useful in determining policy to combat some other public health threats.

10.6 An integrated approach involving individuals, families, social networks, cultures, most government departments and many national organisations is required to successfully challenge obesity.

Learning outcomes for Chapter 10

LO 10.1 Define and use, or recognise definitions and applications of, each of the terms printed in **bold** in the text.

LO 10.2 Describe the factors that create an obesogenic environment and identify measures that might be taken to reduce the levels of obesity in the UK and elsewhere.

LO 10.3 Discuss how ethics and stakeholders can be used to identify policies that might be acceptable in reducing current and future levels of obesity.

LO 10.4 Explain how modelling might inform policy making.

Self-assessment questions for Chapter 10

Question 10.1 (LOs 10.1 and 10.2)

Linda's weight management group encouraged the attendance of family or friends and Sharon went along with Linda. To begin with, Doug didn't like coming home and sometimes finding Linda was out at a class or still at the animal sanctuary and that there was no cake in the house. But after a while he found he was looking forward to her funny stories about the animals and her work there and she definitely became more attractive to him as she lost 35 kg over the following 2 years.

(a) Is this an example of good social marketing or good social capital?

(b) Identify the possible changes that Linda has made to her obesogenic environment that have helped her to lose weight.

Question 10.2 (LOs 10.3 and 10.4)

The effect of manipulations to sales taxes, shown in Figure 10.7, excludes the likely effect on sugar intake, which would be to increase consumption by between 8% and 9% in Social Classes IV and V. The model suggests that this effect could be removed if sugar were taxed at a level that would probably double the price of a bag of sugar. How would you rate this proposal in relation to the criteria identified by the Nuffield Council for Bioethics (Section 10.4.2)? For each criterion, give a rating of inconsistent, consistent or irrelevant.

ANSWERS AND COMMENTS

Answers to self-assessment questions

Question 1.1

The estimate of the cost of obesity is based on a *prediction* of obesity prevalence for 2050, which, in turn, is based in part on statistics for childhood obesity (Section 1.1). There are currently no universally agreed criteria for measuring obesity in childhood (Section 1.3) and published statistics for the prevalence of obesity can change as new information becomes available (Section 1.4.1). Thus 'experts' might disagree on the estimate of the level of obesity in 2050.

Calculating the cost of obesity involves making assumptions about the *relative* importance of obesity as a contributory factor in the prevalence of a suite of comorbidities. In the past, ideas about attributable risk have changed (Section 1.5) so 'experts' might disagree on how much of the costs of treating comorbidities should be attributed to obesity.

There are direct and indirect costs associated with obesity (Section 1.5) and 'experts' might disagree with the categorisation and extent to which these costings are relevant.

You may have thought of other valid reasons.

Question 1.2

The women's BMI is 32, so they are therefore both classified as obese. They are both carrying a lot of weight for their height and this may well affect their joints and lead to osteoarthritis in later years. But Sarah must be physically fit and it is likely that her excess weight is a result of well-developed musculature. She is unlikely to be unhealthy because of her weight.

Question 1.3

There are a number of possibilities here and you may have chosen different examples.

(a) Fat is not stored randomly within the body, but in distinct areas (fat depots) that are characteristic of gender (Section 1.2). (b) The accumulation of fat in specific areas is associated with a high risk of comorbidities, and health professionals have a number of ways of measuring specific adiposity and assessing an individual's health risk (Sections 1.3 and 1.4.2). (c) The cost of treating obesity and the consequences of obesity indicate that the *obesity epidemic* is a public health issue of considerable importance (Section 1.5). (d) Studies show that some groups are at greater risk of becoming obese than others. For example, the children of women who go out to work are more likely to be obese than the children whose mothers do not work (Section 1.6.3).

Question 1.4

From Figure 1.2, it can be seen that the prevalence of obesity is far greater for Mauritanian women than for Mauritanian men. The reason for this difference is that in Mauritania being well nourished is a mark of wealth, and young women are made to overeat so that they will be desirable as brides (Section 1.6.3).

Question 2.1

Ben moves to a state of negative energy balance when he stops driving buses, exercises in the gym and takes up a more physically demanding job. At this point, his energy expenditure exceeds his energy intake and he starts to lose weight.

Question 2.2

Between 12 and 18% of body weight is protein, but it is not a storage material. Rather, it is the fabric of the body, e.g. muscles, skin and other tissues (Section 2.4). Amino acids from the diet can be used to provide energy via metabolic pathways that feed into the TCA cycle (Box 2.1). In starvation, energy can be derived from the breakdown of muscle tissue. This is an abnormal situation and it is not correct, therefore, to describe muscle protein as an energy storage tissue (Section 2.4).

Question 2.3

(a) Partially true. Energy can be stored in any body cell, but only for a few seconds – not for 24 hours (Section 2.6). (b) False. Proteins are made from amino acids (Section 2.4.3). ATP provides the energy necessary for this synthesis (Section 2.6). (c) True. DIT is a measure of the energy input required over 24 hours (Section 2.7). (d) True (Section 2.8.1). Although individuals do not each have a set point for their weight, it appears that they have a settling point and that there are a number of mechanisms that seem to operate as controlling processes in the manner of a homeostatic system.

Question 3.1

Macromolecules are broken down by the teeth chewing food in the mouth, by the stomach muscles churning the chyme and to a limited extent by the muscular peristaltic movements of the gut.

Question 3.2

A lipase inhibitor would prevent the breakdown of lipids in the small intestine. Lipids are not only the storage molecules; there are other fatty molecules in the diet such as cholesterol and the phospholipids. These latter molecules perform several vital roles in the body, so if they were not supplied by the diet there would be very serious health risks. On the other hand, fat is energy-rich, and if the amount of fat that was digested and absorbed was *reduced* rather than totally blocked there would be less chance of TEI exceeding TEE. It must be remembered that both carbohydrate and protein consumed in excess of requirements can be converted to fat for storage (as TAGs), so reducing the intake of dietary fats will not be a total solution to the challenge of obesity.

Question 3.3

Eating a protein-rich diet does not in itself protect against obesity. Proteins are digested to amino acids and used for growth and repair and to maintain the levels of hormones and enzymes. Excess amino acids can be converted to glucose and fatty acids. Just as with the other macronutrients, if TEI exceeds TEE, the surplus is stored as fat.

Question 4.1

Gluco- is short for glucose, *neo-* means new and *genesis* means to make, so the word *gluconeogenesis* means making new glucose. By making new glucose, the body avoids blood sugar levels falling too low during short-term fasting (i.e. when there is a long time between meals). Glucose is produced from non-sugar precursors, namely amino acids (from proteins), lactate (derived from the metabolism of glucose in muscle cells during vigorous exercise) and glycerol (available after TAG breakdown). Gluconeogenesis takes place in liver and kidney cells in response to glucagon activating the enzymes involved in gluconeogenesis.

Question 4.2

The four possible answers are summarised below.

1   Insulin slows glucose production (gluconeogenesis) by the liver. Insulin is circulating when blood glucose levels are high, so it makes sense for insulin to switch off further glucose production. In contrast, glucagon activates the enzymes involved in gluconeogenesis, so the liver returns to making glucose as none is being supplied via ingested food.

2   Insulin increases the storage of glucose as glycogen (glycogenesis) in liver, kidney and muscle cells by stimulating the relevant enzymes. In contrast, glucagon stops the conversion of glucose into glycogen and hence stops glycogen stores being replenished. Glucagon does this by preventing the enzyme glycogen synthase from working (Figure 4.10).

3   Insulin inhibits the release of glucose from glycogen (glycogenolysis) in liver, kidney and muscle cells by its inhibitory effects on the enzymes for this metabolic pathway within the cells. In contrast, during short-term fasting, glucagon raises blood glucose levels by stimulating the breakdown of glycogen stores in the liver and elsewhere. The enzyme that enables glycogen breakdown is glycogen phosphorylase and is activated by glucagon (Figure 4.10).

4   Once the liver (and muscle and kidney) glycogen stores are replenished, insulin stimulates the conversion of excess glucose to fatty acids, via the intermediary molecule acetyl CoA (Figure 4.9a). The fatty acids leave the liver and enter the bloodstream, from where they can be taken up by cells of the adipose tissue and used to manufacture TAGs (i.e. stored as fat), again stimulated by insulin (Figure 4.9b). Note also that insulin suppresses the production of fatty acids from TAGs in adipose tissue. In contrast, glucagon enhances the breakdown of TAGs to glycerol and fatty acids, for further use as an additional energy source.

Question 4.3

Hypoglycaemia refers to a lower than optimum blood glucose level, measured as less than 3 mmol l^{-1}. Hyperglycaemia describes blood glucose levels that are abnormally high.

Question 4.4

Glucose is taken into adipose tissue and converted to TAG when glycogen stores are replenished and blood glucose levels remain high. This conversion takes place in two stages: first, glucose is converted to glycerol; second, glycerol combines with fatty acids to form TAG. These conversions are stimulated by insulin.

Question 4.5

An obese individual has an excess of fat stored. Their swollen adipocytes send out signal molecules such as RBP4 and visfatin. It is believed that such molecules interfere with the normal activity of insulin (Section 4.6.2). Thus, after consuming food, levels of glucose remain high in the blood because insulin is not assisting the entry of glucose into liver, muscle or adipose tissue. This state is known as insulin resistance (Section 4.5). The pancreas will initially respond to the high levels of blood glucose by secreting even more insulin to which the target tissues may respond by taking in a little glucose. However, the continuing high levels of insulin in the blood will lead to downregulation of the receptors for insulin in the target tissues, increasing insulin resistance.

Not all body cells need insulin to help glucose enter. The high levels of glucose in the blood mean that there will be a steep concentration gradient driving more glucose than is required into those cells (e.g. cells in the eye) that are not insulin-dependent. Glycation – the binding of glucose to proteins – causes irreversible structural damage in various cells of the eye.

Question 4.6

Insulin resistance eventually leads to hyperglycaemia, and prolonged exposure of tissue to high blood glucose levels (also called glucose toxicity) causes tissue inflammation and other damage, such as glycation. Glucose toxicity may result in damage to the eyes, red blood cells, nerves in body extremities and kidneys and the development of cardiovascular diseases. High blood glucose levels also provide a nutrient-rich environment for pathogens such as bacteria, and insulin resistance is also thought to lead to the development of diabetes.

Question 4.7

When adipose cells have capacity for further storage, they secrete different amounts of hormonal signals, known as adipokines, such as leptin and adiponectin. Low levels of leptin will lead to enhanced appetite in an attempt to increase adipose reserves. Adiponectin improves insulin sensitivity and promotes glucose uptake from the blood, leading to its conversion and eventual accumulation as fat.

Question 5.1

Hunger describes the state in which eating is likely, whereas satiety describes the state in which further eating is unlikely. Using these two different words emphasises the way in which different mechanisms stimulate and inhibit eating. For example, blood glucose and ghrelin levels may be among the factors that initially stimulate eating. However, sensory-specific satiety, increasing levels of CCK and insulin, as well as increasing energy availability well after a meal, are all likely to be important components of the satiety cascade.

Question 5.2

In addition to showing a correlation between blood levels of the hormone and eating, at least three kinds of observation would be helpful. First, it should be possible to make the synthetic hormone and show that, when it was administered, eating behaviour was stimulated. Second, it should be possible to isolate the receptor for the hormone and show that it was present in parts of the body – most probably the brain – that were directly relevant to eating behaviour. Third, having discovered something about the receptor, it should be possible to synthesise a drug which acts to block the receptor (an antagonist) and show that administration of this substance leads to a reduction in eating behaviour.

Question 5.3

A key point here is that there are two important groups of neurons in this area: those that contain αMSH and those that contain NPY. The former inhibit eating, whereas the latter stimulate eating. Both types of cells receive signals from leptin, ghrelin, insulin and serotonin that help to signal body energy reserves and the consequences of eating. The signals have opposite effects on the two groups of neurons. In addition, these cells – or others to which they are connected – receive inputs from other areas such as the nucleus of the solitary tract. This area will have picked up information relevant to eating from taste neurons and also from the vagus nerve. It is the combination of all these signals that is integration.

Question 5.4

This question can be thought of in a number of different ways. Behavioural observations and experiments described in this chapter have shown how stimuli can, through conditioning, become signals for feeding. The obesogenic environment is one that is full of such signals. Sensory-specific satiety can lead to greatly increased food intake. The obesogenic environment is one in which there is a great variety of food available. Humans evolved in an environment where foods were of low energy density and low in fat. The obesogenic environment is one in which foods have high energy density which is often not very apparent and, although humans may be sensitive to these differences, they may not compensate as fully as they should when they eat energy-dense foods.

Question 6.1

An MC_4 receptor is a membrane protein, found in neurons of the arcuate nucleus in the hypothalamus, that responds to the neurotransmitter αMSH. It is implicated in satiety because its absence results in severe, early onset obesity.

Question 6.2

We each have a different combination of gene variants in our cells. Each gene variant of a particular gene provides slightly different instructions for the same protein as every other gene variant of that gene. These gene variants result in the same protein, but with slightly different characteristics, being faster or slower, more responsive or less responsive, active or inactive. Such differences in gene variants and their protein products are possible for every gene, which gives rise to differences between individuals.

Question 6.3

Generally, the phrase means that the effects of one gene are altered by another gene. More specifically, the phrase means that the effects of a gene variant of one gene differ depending on which gene variant of another gene is present.

Question 6.4

Monogenic obesity refers to the condition where a specific gene variant results in obesity. Genetic susceptibility refers to the likelihood that in a given environment someone will become obese. Someone with a specific gene variant that results in obesity is very likely to become obese and so has a high susceptibility, and that susceptibility is high in many environments. Only one gene is involved in monogenic obesity, whereas many genes contribute to genetic susceptibility.

Question 6.5

They might worry because the daily energy value given in Table 6.2 for a 10-month-old boy is 920 kcal d^{-1}; 800 kcal is a lot less and they might worry that their son is not eating enough. They might not worry for reasons such as: their son is small for his age and hence needs less energy; their son has a cold and is not eating well at the moment; the values in the table are averages and the measurement of energy intake over one day is not representative of general eating patterns.

Question 6.6

The absence of breastfeeding would not contribute to weight gain following pregnancy but it would prevent weight loss. The 170 kcal d^{-1} that would have been mobilised from body tissue for milk production would not be mobilised for that purpose and would either remain or have to be worked off through diet and/ or exercise.

Question 7.1

The effect of considerable amounts of visceral fat is to make it more difficult for the lungs to expand and take in air, so it requires more effort to breathe. If less air is taken in with each breath, there will be less oxygen in the bloodstream. This will be detected by sensors that will send a message to the brain to increase the rate of breathing, resulting in experiencing breathlessness. (This is another homeostatic mechanism; see Figure 2.8.)

Question 7.2

You read that Linda eats large quantities of crisps – far more than most people would manage. BED is often associated with weight cycling, but although Linda tries to lose weight it does not seem that she is successful and so she would not be described as a weight cycler. More information would be needed about Linda's feelings when she binge eats to diagnose her with BED.

Question 7.3

Neither Angus nor Linda have had any *encouragement* to lose weight; Linda's doctor appears unsympathetic, whereas Angus's doctor seems uninterested. Such attitudes are unlikely to make patients wish to lose weight. Nazneen has had health risks explained and has had information to help her modify her eating behaviour. The interest from her midwife may encourage her to lose weight.

Question 7.4

Nazneen may not have realised that she was overweight. Although she had had trouble conceiving, it does not seem as though any health professional has suggested that this could be associated with her weight. She is only given leaflets on healthy eating in pregnancy at 26 weeks when glucose is found in her urine.

Question 8.1

The diets in this chapter differ in their energy content (e.g. low-calorie diets versus VLCDs) or in the balance of their macronutrients (i.e. the proportion of the diet that is composed of protein, carbohydrate and fats).

Question 8.2

One advantage is that resistance exercise can be performed by less mobile individuals. Another advantage is that resistance exercise promotes lean tissue at the expense of adipose tissue. You may also have mentioned that it appears to increase resting metabolic rate, but that is an indirect consequence of its effect on lean tissue.

Question 8.3

You would be annoyed. A 'compensator' would tend to eat more to compensate for energy lost through exercise, making actual weight loss more difficult.

Question 8.4

You might have suggested any number of valid reasons that are not mentioned here, but some of the more likely ones are: the duration of the study (shorter studies usually show greater effects); the persistence of the participants; the sample size (larger samples are preferred); the degree of social support; the way the different groups were established and treated; the exact components of the diet.

Question 8.5

The education campaign would provide explicit knowledge; you would know that eating particular foods was unhealthy. Implicit knowledge could influence behaviour contrary to the education campaign, increasing temptation when you happen to find yourself with the opportunity to have unhealthy foods.

Question 8.6

9.6 g per week is equivalent to 500 g over the year. There are 7700 kcal in each kilogram (1000 g) of body fat, so Paul has a positive energy balance for the year of

$$\frac{500 \text{ g}}{1000 \text{ g}} \times 7700 \text{ kcal} = 3850 \text{ kcal}$$

Question 9.1

Linda has a BMI of just over 40, so she would be a candidate for either drug or surgical treatment on this basis. However, the other NICE guidelines are not met because, although she has tried to lose weight by dieting, she does not appear to have had proper 'dietary and exercise advice' from health professionals. A GP who dismisses her problem with advice to 'eat less and exercise more' is not providing an active weight management strategy for obese patients.

Question 9.2

Randomised double-blind clinical trials involve randomly assigning the participants in the trial to either an active or an inactive drug treatment. Neither the patient nor the treating physician is aware of the condition to which an individual is assigned. The critical feature of such trials is that they control for *placebo* effects.

Question 9.3

Rimonabant is an antagonist at synapses containing cannabinoid receptors. These receptors are found in several brain areas that are important in the regulation of eating, including the hypothalamus and nucleus accumbens. Cannabinoid neurotransmitters are important in signalling the rewarding effects of food, so an antagonist will tend to reduce these effects.

Question 9.4

There are several reasons for the more restrictive conditions for surgical treatment of obesity. Surgical treatments are either difficult or impossible to reverse. The surgical procedure itself has risks, and the general risks of surgery (e.g. anaesthesia) are greater for obese individuals. In addition, the subsequent side effects are more likely to impact on an individual's quality of life.

Question 10.1

(a) This example demonstrates that Linda has good social capital. She has the support of Sharon and the others in the weight management group. It seems that Doug might have been a bit negative to begin with, but he soon started to appreciate the 'new' Linda. Her old friend encouraged her to do some voluntary work, which Linda appears to relish. (b) Going out to work will have increased Linda's activity, and while she is there it will be more difficult to snack. She has obviously made changes to the food she keeps in the house – Doug misses the cakes!

Question 10.2

A tax on sugar would be consistent with criteria 1–4 and 7, whereas 5 and 6 are probably irrelevant. The proposal might be considered to be inconsistent with respect to constraint 9 by some people but probably would be not seen to conflict with constraints 8 and 10.

Comments on activities

Activity 2.1

Here are some thoughts; you may have had other ideas on your list.

Energy intake

Signals from the body that might make you start to eat:

- hunger, rumbling stomach (i.e. a physiological need)
- sight or smell of food (i.e. food availability)
- habit (i.e. a learned behaviour).

Signals from the body that might make you stop eating:

- feeling full (physiological need satisfied)
- seeing an empty plate (food not available)
- deciding you've eaten enough (learned behaviour).

Energy expenditure

This is a tricky one! We weren't *aware* of any body signals to which we would respond to alter BMR or DIT.

In relation to physical activity, the motivation to move would be the necessity of getting from A to B and reaching B would be the signal to stop (unless previously exhausted!).

Activity 4.1

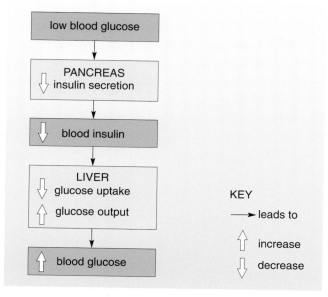

Figure 4.19 Completed Figure 4.7.

The low blood glucose would have a negative effect on insulin secretion, resulting in lowered insulin levels in the blood. This would lead to the liver (and other target tissues) reducing their glucose uptake from the blood. If blood glucose remains low because none is supplied from digested food, the liver would supply glucose into the bloodstream (by undertaking glycogenolysis and gluconeogenesis). The response of the liver would raise blood glucose levels close to normal levels and maintain them there for a few hours, after which another meal would normally be available.

Activity 4.2

In the fed state, the brain receives most of its energy from blood glucose, derived from digested nutrients, to which it has unrestricted access. In the fasted state, blood glucose is not available from digested nutrients; however, the liver releases glucose from its glycogen stores. In addition, TAG stores are broken down to fatty acids and glycerol, with the glycerol converted to glucose via gluconeogenesis. After prolonged fasting, the fatty acids are converted to ketones in the liver and released into the bloodstream. The brain makes use of both glucose and ketones as an energy source during fasting. You may also have described the different roles of the hormones insulin and glucagon in these different physiological states.

Activity 6.1

There are several strategies open to families. However, as with many lifestyle changes, the gradual introduction of changes is recommended:

- reduce portion sizes
- reduce availability of sugar-sweetened drinks, especially between meals
- reduce the availability of energy-dense foods
- promote activity.

These suggestions are similar to those appearing in the *British Medical Journal* (Reilly and Wilson, 2006) namely to:

- ensure that families monitor their own diet, activity, television viewing and computer use
- introduce dietary changes
- reduce sedentary behaviour (particularly television viewing) – to less than 2 hours per day
- increase physical activity through lifestyle changes such as walking to and from school.
- resolve comorbidity (associated disease) if it is present
- achieve weight maintenance not weight loss.

Activity 8.1

Very few people can reliably remember exactly what and how much they have eaten in the previous 24 hours!

Activity 9.1

It depends what you chose for your 'meal' but if wanting a variety of flavours you would be confined to just a few teaspoonfuls of 'ordinary' food (Figure 9.12). Note that once the target weight is achieved it will be necessary to consume a balanced diet, as advised by a dietitian, that provides sufficient calories for weight maintenance.

Figure 9.12   Five 100 ml 'meals'.

In thinking about what you would eat for your three 100 ml meals a day, in addition to quantity of food you would need to think about consuming a diet that provided essential nutrients not stored by the body.

Most energy requirements are to be met by mobilising the fat stored in adipose tissue but some carbohydrate is also necessary. There is also a requirement for some essential amino acids, fatty acids and for vitamins and minerals (Section 2.2). Clearly, it will be necessary to have a diet that has been developed by a dietitian and to adhere to it in order to maintain good health.

Activity 10.1

Here is a list of policies that you may have come up with:

- tax junk foods
- subsidise healthy foods
- remove subsidies from crops that are ingredients of junk foods
- put large health warnings on junk food
- impose a uniform labelling system for foods, highlighting fat and energy content
- provide the least wealthy families with vouchers to use on healthy food
- make school meals compulsory and free for all at primary schools
- increase the time for physical exercise at school
- finance school playing fields and a range of exercise equipment
- ban parking within 400 m of any school
- ban cars from all town and city centres
- impose parking charges on all 'out-of-town' shopping centres
- ensure new housing developments have local shops and parks
- increase the cycle lane network
- restrict planning permission on fast-food outlets
- ensure stairs are more prominent than lifts
- improve footpaths
- ban advertising of foods directed at children
- require GPs to advise about diet and exercise
- provide free membership for all sports clubs.

Activity 10.2

You should find that you have mostly 'no' in relation to the libertarian framework. The balance of 'yes' and 'no' for the utilitarian framework will depend on your judgement of the relative costs and benefits.

REFERENCES AND FURTHER READING

References

Abate, N. and Garg, A. (1995) 'Heterogeneity in adipose tissue metabolism: causes, implications and management of regional adiposity', *Progress in Lipid Research*, vol. 34, pp. 53–70.

Adams, K.F., Schatzkin, A., Harris, T.B., Kipnis, V., Mouw, T., Ballard-Barbash, R., Hollenbeck, A. and Leitzmann, M.F. (2006) 'Overweight, obesity, and mortality in a large prospective cohort of persons 50 to 71 years old', *New England Journal of Medicine*, vol. 355, pp. 763–78.

Ainsworth, B.E. (2002) *The compendium of physical activities tracking guide*, Prevention Research Center, Norman J. Arnold School of Public Health, University of South Carolina [online] Available from: http://prevention.sph.sc.edu/tools/docs/documents_compendium.pdf (Accessed December 2007).

Akobeng, A.K. and Heller, R.F. (2007) 'Assessing the population impact of low rates of breast feeding on asthma, coeliac disease and obesity: the use of a new statistical method', *Archives of Disease in Childhood*, vol. 92, pp. 483–5.

Allender, S. and Rayner, M. (2007) 'The burden of overweight and obesity-related ill health in the UK', *Obesity Reviews*, vol. 8, pp. 467–73.

Baker, J.L., Michaelsen, K.F., Sørensen, T.I. and Rasmussen, K.M. (2007) 'High pre-pregnant body mass index is associated with early termination of full and any breastfeeding in Danish women', *American Journal of Clinical Nutrition*, vol. 86, pp. 404–11.

Banting, W. (1869) *Letter on corpulence, addressed to the public*. Pamphlet [online] Available from: http://www.bantingdiet.com (Accessed May 2008).

Blundell, J.E., Stubbs, R.J., Golding, C., Croden, F., Alam, R., Whybrow, S., Le Noury, J. and Lawton, C.L. (2005) 'Resistance and susceptibility to weight gain: individual variability in response to a high-fat diet', *Physiology & Behavior*, vol. 86, pp. 614–22.

Brandsma, L. (2005) 'Physician and patient attitudes towards obesity', *Eating Disorders*, vol. 13, pp. 201–11.

Campbell, K., Engle, H., Timperio, A., Cooper, C. and Crawford, D. (2000) 'Obesity management: Australian general practitioners' attitudes and practices', *Obesity Research*, vol. 8, pp. 459–66.

Cannon, W.B. and Washburn, A.L. (1912) 'An explanation of hunger', *American Journal of Physiology*, vol. 29, pp. 441–54.

Carryer, J. (2001) 'Embodied largeness: a significant women's health issue', *Nursing Inquiry*, vol. 8, pp. 90–7.

Carpenter, K.M., Hasin, D.S., Allison, D.B. and Faith, M.S. (2000) 'Relationships between obesity and DSM-IV major depressive disorder, suicide ideation, and suicide attempts: results from a general population study', *American Journal of Public Health*, vol. 90, pp. 251–7.

Clifton, P.G., Burton, M.J. and Sharp, C. (1987) 'Rapid loss of stimulus-specific satiety after consumption of a second food', *Appetite*, vol. 9, pp. 149–56.

Cross-Government Obesity Unit, Department of Health and Department for Children, Schools and Families (2008) *Healthy weight, healthy lives: a cross-government strategy for England* [online] Available from: http://www.dh.gov.uk/en/Publichealth/Healthimprovement/Obesity/DH_082383 (Accessed May 2008).

Davenport, C.B. (1923) 'Body build and its inheritance', *Proceedings of the National Academy of Sciences*, vol. 9, pp. 226–30.

Demerath, E.W., Choh, A.C., Czerwinski, S.A., Lee, M., Sun, S.S., Chumlea, W.C., Duren, D., Sherwood, R.J., Blangero, J., Towne, B. and Siervogel, R.M. (2007) 'Genetic and environmental influences on infant weight and weight change: the Fels Longitudinal Study', *American Journal of Human Biology*, vol. 19, pp. 692–702.

Department of Health (DoH) (1991) *Dietary reference values for food energy and nutrients in the United Kingdom*, Report on health and social subjects 41: Report of the panel on dietary reference values of the committee on medical aspects of food policy, London, HMSO.

Dubois, L., Farmer, A., Girard, M. and Peterson, K. (2007) 'Regular sugar-sweetened beverage consumption between meals increases risk of overweight among preschool-aged children', *Journal of the American Dietetic Association*, vol. 107, pp. 924–34.

Dulloo, A.G. and Miller, D.S. (1987) 'Obesity: a disorder of the sympathetic nervous system', *World Review of Nutrition and Dietetics*, vol. 50, pp. 1–56.

Edmunds, L.D. (2005) 'Parents' perceptions of health professionals' responses when seeking help for their overweight children', *Family Practice*, vol. 22, pp. 287–92.

Epstein, L.H., Valoski, A.M., Vara, L.S., McCurley, J., Wisniewski, L., Kalarchian, M.A., Klein, K.R. and Shrager, L.R. (1995) 'Effects of decreasing sedentary behavior and increasing activity on weight change in obese children', *Health Psychology*, vol. 14, pp. 109–15.

Faith, M.S., Scanlon, K.S., Birch, L.L., Francis, L.A. and Sherry, B. (2004) 'Parent–child feeding strategies and their relationships to child eating and weight status', *Obesity Research*, vol. 12, pp. 1711–22.

Field, A.E., Manson, J.E., Taylor, C.B., Willett, W.C. and Colditz, G.A. (2004) 'Association of weight change, weight control practices, and weight cycling among women in the Nurses Health Study II', *International Journal of Obesity*, vol. 28, pp. 1134–42.

Fisher, J.O. (2007) 'Effects of age on children's intake of large and self-selected food portions', *Obesity*, vol. 15, pp. 403–12.

Flegal, K.M., Graubard, B.I., Williamson, D.F. and Gail, M.H. (2007) 'Cause-specific excess deaths associated with underweight, overweight, and obesity', *Journal of the American Medical Association*, vol. 298, pp. 2028–37.

Fontaine, K.R., Faith, M.S., Allison, D.B. and Cheskin, L.J. (1998) 'Body weight and health care among women in the general population', *Archives of Family Medicine*, vol. 7, pp. 381–4.

Foster, G.D., Wadden, T.A., Makris, A.P., Davidson, D., Sanderson, R., Allison, D.B. and Kessler, A. (2003) 'Primary care physicians' attitudes about obesity and its treatment', *Obesity Research*, vol. 11, pp. 1168–77.

French, J. and Blair-Stevens, C. (2005) *Social Marketing Big Pocket Guide* [online] Available from: http://www.nsms.org.uk/images/CoreFiles/NSMC_Big_Pocket_Guide_Aug_2007.pdf (Accessed May 2008).

Galuska, D.A., Will, J.C., Serdula, M.K. and Ford, E.S. (1999) 'Are health care professionals advising obese patients to lose weight?', *Journal of the American Medical Association*, vol. 282, pp. 1576–8.

Gardner, C.D., Kiazand, A., Alhassan, S., Kim, S., Stafford, R.S., Balise, R.R., Kraemer, H.C. and King, A.C. (2007) 'Atkins, Zone, Ornish and LEARN – diets for change in weight and related risk factors among overweight premenopausal women: the A to Z Weight Loss Study: a randomised trial', *Journal of the American Medical Association*, vol. 297, pp. 969–77.

Grilo, C.M., Masheb, R.M., Brody, M., Toth, C., Burke-Martindale, C.H. and Rothschild, B.S. (2005) 'Childhood maltreatment in extremely obese male and female bariatric surgery candidates', *Obesity Research*, vol. 13, pp. 123–30.

Haines, J., Neumark-Sztainer, D., Eisenberg, M.E. and Hannan, P.J. (2006) 'Weight teasing and disordered eating behaviours in adolescents: longitudinal findings from project EAT (Eating Among Teens)', *Pediatrics*, vol. 117, pp. 209–15.

Hansen, D.L., Toubro, S., Stock, M.J., Macdonald, I.A. and Astrup, A. (1998) 'Thermogenic effects of sibutramine in humans', *American Journal of Clinical Nutrition*, vol. 68, pp. 1180–6.

Hansen, D.L., Toubro, S., Stock, M.J., Macdonald, I.A. and Astrup, A. (1999) 'The effect of sibutramine on energy expenditure and appetite during chronic treatment without dietary restriction', *International Journal of Obesity and Related Metabolic Disorders*, vol. 23, pp. 1016–24.

Haslam, D. (2007) 'Obesity: a medical history', *Obesity Reviews*, vol. 8, pp. 31–6.

Haugen, F. and Drevon, C.A. (2007) 'The interplay between nutrients and the adipose tissue', *Proceedings of the Nutrition Society*, vol. 66, pp. 171–82.

Heslehurst, N., Lang, R., Rankin, J., Wilkinson, J. R. and Summerbell, C.D. (2007) 'Obesity in pregnancy: a study of the impact of maternal obesity on NHS maternity services', *BJOG*, vol. 114, pp. 334–42.

House of Commons Health Committee (2004) *Obesity: third report of session 2003–04*, vol. 1, London, The Stationery Office.

Jakicic, J.M., Marcus, B.H., Gallagher, K.L., Napolitano, M. and Lang, W. (2003) 'Effect of exercise duration and intensity on weight loss in overweight, sedentary women: a randomized trial', *Journal of the American Medical Association*, vol. 290, pp. 1323–30.

James, W. (1890) *The Principles of Psychology* [online] Available from: http://psychclassics.yorku.ca/James/Principles/index.htm (Accessed January 2008).

Jebb, S.A. (2007) 'Dietary determinants of obesity', *Obesity Reviews*, vol. 8, pp. 93–7.

Jeffery, R.W., Wing, R.R., Sherwood, N.E. and Tate, D.F. (2003) 'Physical activity and weight loss: does prescribing higher physical activity goals improve outcome?', *American Journal of Clinical Nutrition*, vol. 78, pp. 684–9.

King, N.A., Caudwell, P., Hopkins, M., Byrne, N.M., Colley, R., Hills, A.P., Stubbs, J.R. and Blundell, J.E. (2007) 'Metabolic and behavioural compensatory responses to exercise interventions: barriers to weight loss', *Obesity*, vol. 15, pp. 1373–83.

Kral, T.V. and Rolls, B.J. (2004) 'Energy density and portion size: their independent and combined effects on energy intake', *Physiology & Behavior*, vol. 82, pp. 131–8.

Lobstein, T. and Jackson Leach, R. (2007) *Tackling Obesities: future choices*, Long Science Review [online] Available from: http://www.foresight.gov.uk/Obesity/Obesity_final/Index.html (Accessed January 2008).

Loos, R.J. and Bouchard, C. (2008) '*FTO*: The first gene contributing to common forms of human obesity', *Obesity Reviews*, vol. 9, pp. 246–50.

Luszczynska, A., Sobczyk, A. and Abraham, C. (2007) 'Planning to lose weight: randomized controlled trial of an implementation intention prompt to enhance weight reduction among overweight and obese women', *Health Psychology*, vol. 26, pp. 507–12.

Ma, L.P., Tataranni, A., Bogardus, C. and Baier, L.J. (2004) 'Melanocortin 4 receptor gene variation is associated with severe obesity in Pima Indians', *Diabetes*, vol. 53, pp. 2696–9.

Maroney, D. and Golub, S. (1992) 'Nurses' attitudes toward obese persons and certain ethnic groups', *Perceptual and Motor Skills*, vol. 75, pp. 387–91.

McCabe, C. and Rolls, E.T. (2007) 'Umami: a delicious flavor formed by convergence of taste and olfactory pathways in the human brain', *European Journal of Neuroscience*, vol. 25, pp. 1855–64.

McElroy, S.L., Kotwal, R., Malhotra, S., Nelson, E.B., Keck, P.E. and Nemeroff, C.B. (2004) 'Are mood disorders and obesity related? A review for the mental health professional', *Journal of Clinical Psychiatry*, vol. 65, pp. 634–51.

Mill, J.S. (1859) *On Liberty* [online] Available from: http://www.constitution.org/jsm/liberty.htm (Accessed July 2008).

Millstone, E. and Lobstein, T. (2007) 'The PorGrow project: an introduction and overview', *Obesity Reviews*, vol. 8, suppl. 2, pp. 5–6.

Nassis, G., Papantakou, K., Skenderi, K., Triandafillopoulou, M., Kavouras, S., Yannakoulia, M., Chrousos, G. and Sidossis, L. (2005) 'Aerobic exercise training improves insulin sensitivity without changes in body weight, body fat, adiponectin, and inflammatory markers in overweight and obese girls', *Metabolism*, vol. 54, pp. 1472–9.

National Institute for Health and Clinical Excellence (NICE) (2006) *CG43 Obesity: Full guideline* [online] Available from: http://www.nice.org.uk/guidance/CG43/guidance (Accessed May 2008).

Neovius, M., Linne, Y., Barkeling, B. and Rössner, S. (2004) 'Discrepancies between classification systems of childhood obesity', *Obesity Reviews*, vol. 5, pp. 105–14.

Ness, A.R., Leary, S.D., Mattocks, C., Blair, S.N., Reilly, J.J., Wells, J., Ingle, S., Tilling, K., Smith, G.D. and Riddoch, C. (2007) 'Objectively measured physical activity and fat mass in a large cohort of children', *PLoS Medicine*, vol. 4, e97.

Ochoa, M.C., Marti, A., Azcona, C., Chueca, M., Oyarzábal, M., Pelach, R., Patiño, A., Moreno-Aliaga, M.J., Martínez-González, M.A. and Martínez, J.A. (2004) 'Gene–gene interaction between PPAR gamma 2 and ADR beta 3 increases obesity risk in children and adolescents', *International Journal of Obesity and Related Metabolic Disorders*, vol. 28, suppl. 3, pp. S37–41.

Oken, E., Taveras, E.M., Kleinman, K.P., Rich-Edwards, J.W. and Gillman, M.W. (2007) 'Gestational weight gain and child adiposity at age 3 years', *American Journal of Obstetrics and Gynecology*, vol. 196, pp. 322.e1–8.

Ong, K.K.L., Ahmed, M.L., Emmett, P.M., Preece, M.A. and Dunger, D.B. (2000) 'Association between postnatal catch-up growth and obesity in childhood: prospective cohort study', *British Medical Journal*, vol. 320, pp. 967–71.

Paeratakul, S., White, M.A., Williamson, D.A., Ryan, D.H. and Bray, G.A. (2002) 'Sex, race/ethnicity, socioeconomic status, and BMI in relation to self-perception of overweight', *Obesity Research*, vol. 10, pp. 345–50.

Poehlman, E.T., Denino, W.F., Beckett, T., Kinaman, K.A., Dionne, I.J., Dvorak, R. and Ades, P.A. (2002) 'Effects of endurance and resistance training on total daily energy expenditure in young women: a controlled randomized trial', *Journal of Clinical Endocrinology and Metabolism*, vol. 87, pp. 1004–9.

Poppitt, S.D. and Prentice, A.M. (1996) 'Energy density and its role in the control of food intake: evidence from metabolic and community studies', *Appetite*, vol. 26, pp. 153–74.

Powell, K. (2007) 'Obesity: the two faces of fat', *Nature*, vol. 447, pp. 525–7.

Prentice, A.M. and Jebb, S.A. (2003) 'Fast foods, energy density and obesity: a possible mechanistic link', *Obesity Reviews*, vol. 4, pp. 187–94.

Puhl, R. and Brownell, K. (2001) 'Bias, discrimination, and obesity', *Obesity Research*, vol. 9, pp. 788–805.

Puhl, R., Henderson, K.E. and Brownell, K.D. (2005) 'Social consequences of obesity', in Kopelman, P.G., Caterson, I. D. and Dietz, W.H. (eds), *Clinical Obesity in Adults and Children*, Oxford, Blackwell Publishing.

Redman, L.M., Heilbronn, L.K., Martin, C.K., Alfonso, A., Smith, S.R. and Ravussin, E. (2007) 'Effect of calorie restriction with or without exercise on body composition and fat distribution', *Journal of Clinical Endocrinology and Metabolism*, vol. 92, pp. 865–72.

Reilly, J.J. and Wilson, D. (2006) 'ABC of obesity – Childhood obesity', *British Medical Journal*, vol. 333, pp. 1207–10.

Rolls, B.J., Van Duijvenvoorde, P.M. and Rolls, E.T. (1984) 'Pleasantness changes and food intake in a varied four-course meal', *Appetite*, vol. 5, pp. 337–48.

Ross, R. and Janssen, I. (2001) 'Physical activity, total and regional obesity: dose–response considerations', *Medicine and Science in Sports and Exercise*, vol. 33 (suppl.), S521–7.

Schaller, G.B. (1972) *The Serengeti Lion*, Chicago, University of Chicago Press.

Schulz, L.O., Bennett, P.H., Ravussin, E., Kidd, J.R., Kidd, K.K., Esparza, J. and Valencia, M.E. (2006) 'Effects of traditional and western environments on prevalence of type 2 diabetes in Pima Indians in Mexico and the U.S.', *Diabetes Care*, vol. 29, pp. 1866–71.

Schwartz, M.B. and Brownell, K.D. (2004) 'Obesity and body image', *Body Image*, vol. 1, pp. 43–56.

Schwartz, M.B., O'Neal Chambliss, H., Brownell, K.D., Blair, S.N. and Billington, C. (2003) 'Weight bias among health professionals specialising in obesity', *Obesity Research*, vol. 11, pp. 1033–9.

Seidell, J.C. (2005) 'Epidemiology – definition and classification of obesity', in Kopelman, P.G., Caterson, I.D. and Dietz, W.H. (eds), *Clinical Obesity in Adults and Children*, Oxford, Blackwell Publishing.

Sjöström, L., Lindroos, A.K., Peltonen, M., Torgerson, J., Bouchard, C., Carlsson, B., Dahlgren, S., Larsson, B., Narbro, K., Sjöström, C.D., Sullivan, M. and Wedel, H., the Swedish Obese Subjects Study (2004) 'Lifestyle, diabetes, and cardiovascular risk factors 10 years after bariatric surgery', *New England Journal of Medicine*, vol. 351, pp. 2683–93.

Sjöström, L., Narbro, K., Sjöström, C.D., Karason, K., Larsson, B., Wedel, H., Lystig, T., Sullivan, M., Bouchard, C., Carlsson, B., Bengtsson, C., Dahlgren, S., Gummesson, A., Jacobson, P., Karlsson, J., Lindroos, A.K., Lönroth, H., Näslund, I., Olbers, T., Stenlöf, K., Torgerson, J., Agren, G. and Carlsson, L.M., the Swedish Obese Subjects Study (2007) 'Effects of bariatric surgery on mortality in Swedish obese subjects', *New England Journal of Medicine*, vol. 354, pp. 741–52.

Sørensen, T.I., Price, R.A., Stunkard, A.J. and Schulsinger, F. (1989) 'Genetics of obesity in adult adoptees and their biological siblings', *British Medical Journal*, vol. 298, pp. 87–90.

Steffen, R., Biertho, L., Ricklin, T., Piec, G. and Horber, F.F. (2003) 'Laparoscopic Swedish adjustable gastric banding: a five-year prospective study', *Obesity Surgery*, vol. 13, pp. 404–11.

Strauss, R.S. (2000) 'Childhood obesity and self-esteem', *Pediatrics*, vol. 105, p. 15.

Teitelbaum, T. (1964) 'Psychology: a behavioral reinterpretation', *Proceedings of the American Philosophical Society*, vol. 108, pp. 464–72.

Tsai, A.G. and Wadden, T.A. (2006) 'The evolution of very-low-calorie diets: An update and meta-analysis', *Obesity*, vol. 14, pp. 1283–93.

Wang, G.J., Volkow, N.D., Logan, J., Pappas, N.R., Wong, C.T., Zhu, W., Netusil, N. and Fowler, J.S. (2001) 'Brain dopamine and obesity', *The Lancet*, vol. 357, pp. 354–7.

Wardle, J. and Cooke, L. (2005) 'The impact of obesity on psychological wellbeing', *Best Practice and Research Clinical Endocrinology and Metabolism*, vol. 19, pp. 421–40.

Wardle, J., Carnell, S., Haworth, C.M.A. and Plomin, R. (2008) 'Evidence for a strong genetic influence on childhood adiposity despite the force of the obesogenic environment', *American Journal of Clinical Nutrition*, vol. 87, pp. 398–404.

Watkins, P. (2003) 'Cardiovascular disease, hypertension and lipids', *British Medical Journal*, vol. 326, pp. 874–6.

WHO (World Health Organization) (2002) *Reducing risks, promoting healthy life.* Geneva, WHO.

WHO (World Health Organization) (2003) *Diet, nutrition and the prevention of chronic diseases: Report of a joint WHO/FAO Expert consultation*, Geneva, WHO.

Yang, Q., Graham, T.E., Mody, N., Preitner, F., Peroni, O.D., Zabolotny, J.M., Kotani, K., Quadro, L. and Kahn, B.B. (2005) 'Serum retinol binding protein 4 contributes to insulin resistance in obesity and type 2 diabetes', *Nature*, vol. 436, pp. 356–62.

Yeomans, M.R. (1996) 'Palatability and the micro-structure of feeding in humans: the appetizer effect', *Appetite*, vol. 27, pp. 119–33.

Yeomans, M.R., Weinberg, L. and James, S. (2005) 'Effects of palatability and learned satiety on energy density influences on breakfast intake in humans', *Physiology & Behavior*, vol. 86, pp. 487–99.

Zelissen, P.M., Stenlof, K., Lean, M.E., Fogteloo, J., Keulen, E.T., Wilding, J., Finer, N., Rössner, S., Lawrence, E., Fletcher, C. and McCamish, M. (2005) 'Effect of three treatment schedules of recombinant methionyl human leptin on body weight in obese adults: a randomized, placebo-controlled trial', *Diabetes, Obesity and Metabolism*, vol. 7, pp. 755–61.

ACKNOWLEDGEMENTS

Grateful acknowledgement is made to the following sources for permission to reproduce material in this book.

Cover Photo

Daniel Hughes/Fotolia.

Figures

Figure 1.6c: *Bacchus*, 1636–40 (oil on canvas transferred from panel), Rubens, Peter Paul (1577–1640)/Hermitage, St Petersburg, Russia/The Bridgeman Art Library; Figure 1.8: Athletics Northern Ireland; Figure 1.9a: Ian Hooton/ Science Photo Library; Figure 1.9b: © Medical-on-Line/Alamy; Figure 1.11: www.dietandfitness.co.uk; Figure 1.12: Kopelman, P.G. et al. (2005) *Clinical Obesity in Adults and Children*, 2nd edn, Blackwell Publishing; Figure 1.13: © Visual Arts Library (London)/Alamy; Figure 1.14: Adams, K.F. et al. (2006) 'Multivariate relative risks of death in relation to BMI among men', *New England Journal of Medicine*, vol. 355, no. 8, Massachusetts Medical Society; Figure 1.15: Wilkinson, J.R. et al. (2007) 'Surveillance and Monitoring', *Obesity Reviews*, vol. 8, suppl. 1, The International Association for the Study of Obesity; Figures 1.16, 4.15, 5.9 and 10.2: Kopelman P.G. et al. (2005) *Clinical Obesity in Adults and Children*, Blackwell Publishing; Figure 1.19: Lobstein T. (2002) *Obesity – International Comparisons URN 07/926A*, Crown copyright material is reproduced under Class Licence Number C01W0000065 with the permission of the Controller of HMSO and the Queen's Printer for Scotland; Figure 1.20: Getty Images; Figures 1.21 and 10.9: Lobstein, T. and Jackson Leach, R. (2007) *Obesity Future Choices*, Long Science Review; Figure 2.1: Balance of Good Health, reproduced by permission of the Health Education Authority; Figure 2.7: Tomasz Przechlewski/Flickr Photo Sharing; Figure 2.10: Source unknown; Figure 3.2: Academy of Radiology, Norfolk and Norwich University Hospital, Norwich; Figure 3.9: Dr R. Dourmashkin/Science Photo Library; Figure 4.1: Impact Photos; Figure 4.4 Blackwell Science Ltd; Figure 4.14: Caroline Pond, Open University; Figure 4.17: Lee, H.J. and Shoelson, S. in Powell, K. (2007) 'Two faces of fat', *Nature*, vol. 447, Nature Publishing Group; Figure 4.18: Carlson, N.R. (2001) *Physiology of Behaviour*, 7th edn, Pearson Education Inc.; Figure 5.1: www.benainslie.com; Figure 5.4: Based on Logue, A.W. (1991) *The Psychology of Eating and Drinking*, 2nd edn, WH Freeman; Figure 5.5: Arthur, L. et al. (2003) 'Blood Glucose Dynamics and Control and Meal Initiation; a pattern detection system and recognition theory', *Physiological Review*, American Physiological Society; Figure 5.7: Dummings, D.E. et al. (2001) 'A preprandial rise in plasma ghrelin levels suggests a role in meal initiation in humans', *Diabetes*, American Diabetes Association; Figure 5.8: Kirkham, T. C. and Cooper, S. J. (2007) *Appetite and Body Weight*, Elsevier Science; Figure 5.11: Yeomans M., et al., (2005) *Physiology & Behaviour*, Elsevier Science; Figure 5.14: McCabe, C. and Rolls, E.T. (2007) 'Umami: a delicious flavor formed by convergence of taste and Olfactory pathways in the human brain', *European Journal of Neuroscience*, vol. 25, used with permission of the

of Obesity; Figure 10.8: Institute for Agriculture and Trade Policy; Figure 10.10: Health Education Authority; Figure 10.11: © 2008 The Prince's Foundation for Integrated Health.

Tables

Table 7.3: Stunkard, A.J. and Wadden, T.A. (1992) 'Psychological aspects of severe obesity', *American Journal of Clinical Nutrition*, vol. 55, American Society for Clinical Nutrition.

Every effort has been made to contact copyright holders. If any have been inadvertently overlooked the publishers will be pleased to make the necessary arrangements at the first opportunity.

INDEX

Entries and page numbers in **bold type** refer to key words which are printed in **bold** in the text. Indexed information on pages indicated by *italics* is carried mainly or wholly in a figure, table or box.

prevalence of obesity *4*
see also UK
excretion 64
exercise *see* physical activity; sports
explicit 183–4
expressed 38, 67, 118
extracellular fluids 64

F

facilitated diffusion 65, 66, 71
faeces 51, 53, 54, 190
families 120–1, 184
see also parents
'fast food' 29, 114
outlets 47, *48*
portion distortion *76*
see also junk food
fasting 72, 73
metabolic pathways *90*
role of glucagon 80–3
fat body mass 42
fat depots 6–9, 13, 79
in pregnancy *134, 135*
sex differences and 87
signalling molecules and 88–9
'fat suit' *177*
fatness *see* body fat
fats *33, 34,* 35
absorption of 63
energy yields of *39*
metabolic pathways 40, *41*
orlistat and 188, 190
satiating capacity 110–11
see also **fatty acids**; **lipids**;
macronutrients
fatty acids 36, *41,* 54, *54,* 59
absorption 62, 63, *65, 66*
formation of *79*
interconversions *82*
metabolism 144, 170–1
see also fats; **lipids**
feedback loops 76–80
negative feedback 45, 59, *73, 78*
feedforward 46, 58
feeding behaviour *see* eating behaviour
'feeding centre' 94, 112
fenfluramine 196, 197, 199
fertility 149
fibre 34, 53, 54
Finland 223

5-HT$_{2C}$ receptor 197, 199
flavours 104–6
flow diagrams 77–8
food
choices 131
diaries 172
digestion and absorption 51, 60–7
energy content of 29, 39
energy from 31, 32, 39–42
passage through the gut 53–60
prices *226*
quantities eaten *69*
ready availability of 47, 48
satiating capacity 109–11
storage 86
see also balanced diet; 'fast food';
junk food; meals
Food Standards Agency 32–3
formula milk 129
free school meals 26
fruit *33, 34,* 54, 168
FTO gene 123, 126

G

gall bladder 55
gall bladder disease *18,* 148
gangrene 145
gastric banding 204–5, 207
gastric bypass surgery *205, 206*–7
gastric juices 57–8
gastrointestinal system *see* digestive
system
gender 23, *24, 25, 26*
and BMR 42
and body image 155, 156, *157*
and energy requirements 167–8
gene–environment interactions and
127–8
and weight attitudes 159
see also men; women
gene variants 118–19, 123, 124–6
in the Pima Indians 127
**gene–environment interaction 124,
125**–8
gene–gene interaction 126
general practitioners *21,* 32, 152–3
genes
and BMI 121
candidate genes 127

and environment 117, 118
making proteins *37, 38,* 67, 191
and obesity 118–23
see also **mutation**
genetic change, rates of *123*
genetic susceptibility 118, 127, 128
genome 123
gestation 128
gestational diabetes 145, 146, 150
ghrelin 99–100, 113
antagonists 199–200
GLP-1 (glucagon-like peptide) 55, 59,
107–8
glucagon 62–3, 76–7
and fatty acid synthesis 83
role in fasting 80–3
glucagon-like peptide (GLP-1) 55, 59,
107–8
gluconeogenesis 72, 74, 81, *82,* 85
glucose *35, 36, 41*
cellular respiration 39–40
conversion to glycogen *63*
impaired tolerance 146
interconversions *82*
oral glucose tolerance test 146
storage *63,* 72, 73
transporters 71–2, 78
uptake 67, 73
see also blood glucose levels
'glucostatic' hypothesis 97
glycation 70, 85, 145
glycerol 36, *41,* 54, 59, 62
absorption *65, 66*
forming TAGs 79
glycogen 35, 36, *63*
and blood glucose levels 72, 74–5,
77, 79
breakdown in liver 80, *81*
energy reserves *87*
glycogen phosphorylase *79, 80*
glycogen synthase 78
glycogenesis 74, *79*
glycogenolysis 74, 79
government policy 22, 131, 161,
214–15
effectiveness of 222–30
evaluation 215–22
greatly increased relative risk *18*